# Upstream Engineering and Flow Assurance Forum 2017



Core Programming Area at the 2017 AIChE Annual Meeting

Minneapolis, Minnesota, USA
29 October - 3 November 2017

ISBN: 978-1-5108-5811-4

**Printed from e-media with permission by:**

Curran Associates, Inc.
57 Morehouse Lane
Red Hook, NY 12571

CURRAN ASSOCIATES INC.
proceedings.com

**Some format issues inherent in the e-media version may also appear in this print version.**



Printed by Curran Associates, Inc. (2018)



AIChE
120 Wall Street, FL 23
New York, NY 10005-4020

Phone: (800) 242-4363
Fax: (203) 775-5177

www.aiche.org

**Additional copies of this publication are available from:**

Curran Associates, Inc.
57 Morehouse Lane
Red Hook, NY 12571 USA
Phone: 845-758-0400
Fax: 845-758-2633
Email: curran@proceedings.com
Web: www.proceedings.com

# TABLE OF CONTENTS

**Some abstracts in this book refer to images that were not made available by the publisher**

# 498181 Investigation of the Impact of Inorganic Solids on Asphaltenes Precipitation

**Abstract:**

The presence of inorganic solids in Athabasca bitumen was recently reported to significantly increase the rate of asphaltene precipitation in diluted bitumen. The previous research indicated that asphaltenes precipitate on the surface of inorganic solids following a heterogeneous precipitation mechanism. In order to better understand the role of inorganic solids on the solubility and kinetic precipitation of asphaltenes, it is important to study under which conditions the heterogeneous or homogeneous mechanism dominates the precipitation process. In this talk we will explore the impact of inorganic solid and diluent type on the mechanism of asphaltene precipitation. To accomplish this task, inorganic solids were removed from a diluted bitumen, and different kinds of the inorganic solids were added back to study the impact of solid type and concentration on the precipitation kinetics of asphaltenes. The mass of precipitated asphaltenes was measured as a function of time. Moreover, by changing the solid:asphaltene ratio, the influences of homogeneous aggregation and heterogeneous nucleation can be determined. Furthermore, scanning electron microscopy (SEM) with focused ion beam (FIB) milling was used to characterize the precipitated asphaltenes in the presence of inorganic solids to provide visual evidence of the precipitation mechanism. Overall, this study can establish an improved understanding of the influence of inorganic solids on the precipitation kinetics and mechanism of asphaltenes precipitation.

1

# Microstructure of Asphaltenes in Solvent Blends Investigated by Viscometry

Weiyi Kong, Amanda Chew, Michael P. Hoepfner

Department of Chemical Engineering, University of Utah, Salt Lake City, Utah 84112, United States

2

## Introduction

- Asphaltenes are a class of material in crude oil that can destabilize when temperature, pressure or solution components changes.
- Asphaltenes solution is a polydispersity system.
  - Size
  - Sub-structure
  - Properties
- Finding a method to characterize all of asphaltenes is important for isolating these “troublemaking” asphaltenes.

# Background: Polydispersity

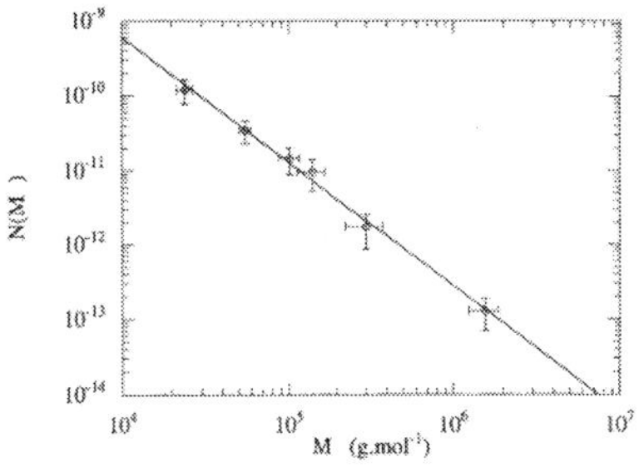


Asphaltenes is polydisperse system

- M and N are power law distribution.

$$N = A \times M^{-a}$$

**N** number of each molecular weight clusters

**A** polydisperse parameter

**M** molecular weight

**a** polydisperse index

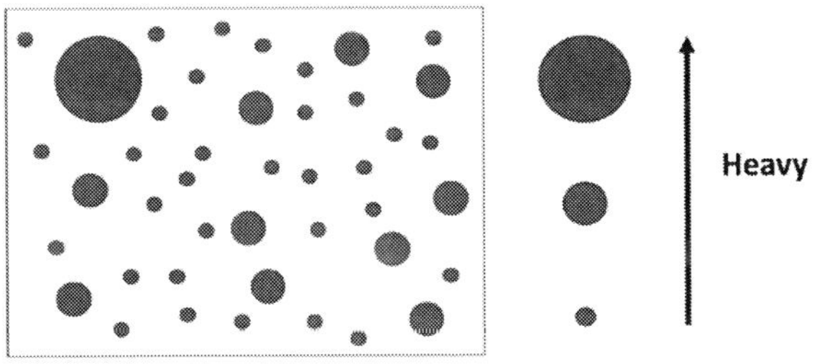


Polydisperse system

[1]Fenistein D, Barre L. Experimental measurement of the mass distribution of petroleum asphaltene aggregates using ultracentrifugation and small-angle X-ray scattering[J]. Fuel, 2001, 80(2): 283-287



# Background: Fractal Dimension ($D_f$)

- A number that quantitatively describes how an object fills its space.[1]
- Small changes in $D_f$ cause large changes in structure

Fractal Example: Colloidal Gold

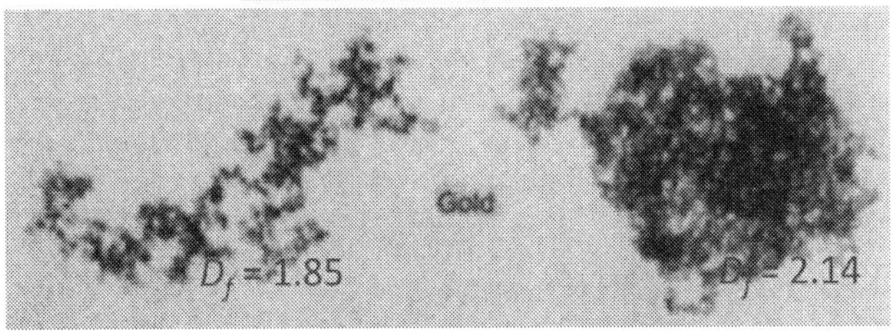


[1] Sagan, Hans (1994). Space-Filling Curves. Berlin: Springer-Verlag. p. 156. ISBN 0-387-94265-3.
[2] Lin, et al., Nature, v.339 (1989)

# Motivation

When adding heptane to an asphaltene solution,

- **Small angle neutron scattering(SANS):** Asphaltene clusters grow larger.
- **Rheometer:** Asphaltene clusters become smaller first then bigger.
- The reliable results are **inconsistent**.
- An opportunity exists for new fundamental discovery.

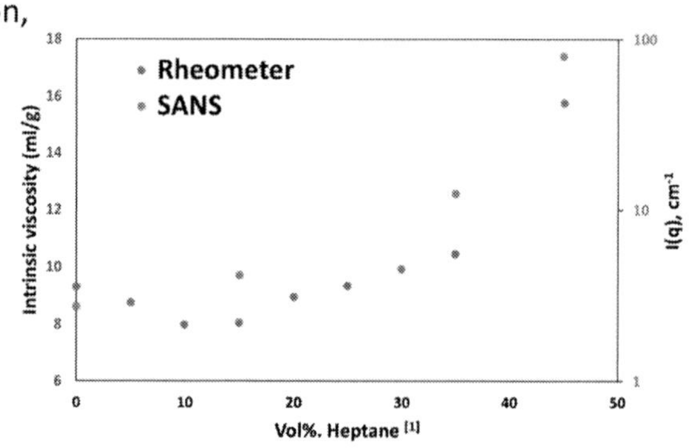


[1]Fenistein D, Barré L, Broseta D, et al. Viscosimetric and neutron scattering study of asphaltene aggregates in mixed toluene/heptane solvents[J]. Langmuir, 1998, 14(5): 1013-1020.



# Hypothesis

**When adding heptane into asphaltenes toluene solution, 3 things will happen:**

1. Asphaltenes aggregate together forming large clusters.
2. Parts of asphaltenes precipitated (insoluble asphaltenes).
3. Insoluble asphaltenes shrink. $D_f$ increases

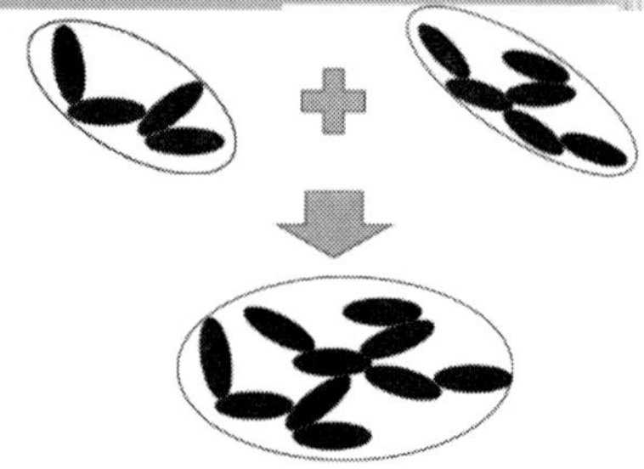

Asphaltenes aggregate

Soluble asphaltenes

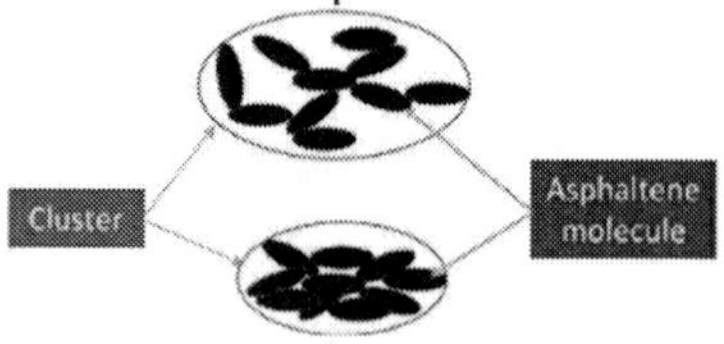


Insoluble asphaltenes

# Model

**Goal : Build a model to explore the effect of polydispersity on asphaltenes intrinsic viscosity.**

**Variables:**

**Adding heptane to an asphaltene solution** → Solubility / Fractal dimension ($D_f$) / Molecular weight distribution → **Intrinsic viscosity**

**Solubility**
Part of soluble asphaltenes become insoluble asphaltenes.

**Fractal dimension ($D_f$)**
Soluble asphaltenes: **low** fractal dimension
Insoluble asphaltenes: **high** fractal dimension

**Molecular weight distribution**
Asphaltene clusters tend to aggregate forming heavier cluster.



# Model

- Fractal dimension (**$D_f$**)
- Molecular weight (**M**)
- Mark-Houwink Eqn

$$[\eta]_i = KM_i^{\frac{3}{Df}-1} \text{ [1]}$$

- Intrinsic viscosity of each size cluster (**$[\eta]_i$**)
- Molecular weight distribution (Number of molecules (**N**))

$$[\eta] = \frac{\sum M_i N_i [\eta]_i}{\sum N_i M_i} \text{ [2]}$$

- The average intrinsic viscosity (**$[\eta]$**)

Result

**Variables & Parameters:**
1. **Fractal dimension ($D_f$)**
2. **Molecular weight (M)**
3. **Mark-Houwink parameter (K)**
4. **Solubility**
5. **Number of each size asphaltenes cluster(N)**


[1] Barré, L., Simon, S., & Palermo, T. (2008). Solution properties of asphaltenes. Langmuir, 24(8), 3709-3717.
[2]Fenistein, Denis, et al. "Viscosimetric and neutron scattering study of asphaltene aggregates in mixed toluene/heptane solvents." Langmuir 14.5 (1998): 1013-1020.

# Model: Parameters and Variables

- **Fractal dimension ($D_f$):** Soluble asphaltenes, $D_f$ is around 1.8.
  Insoluble asphaltenes, $D_f$ is about 2.5.[1]
- **Molecular weight (M):** Soluble asphaltenes from 20000 to 10^6 g/mol; Insoluble asphaltenes is 10^8 g/mol.[2]
- **Mark-Houwink parameter (K):** 0.049[3]
- **Solubility:**
  - Asphaltenes solubility in crude oil.
  - Asphaltenes mass got at 0 vol% heptane
  - Calculate the insoluble asphaltenes ratio

Solubility [1]

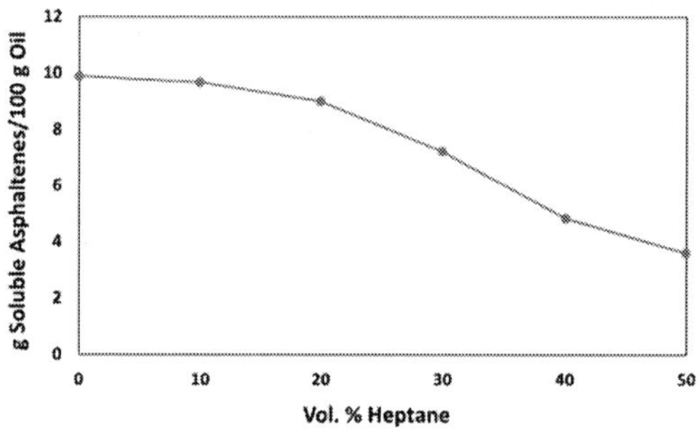



[1]Hoepfner M P, Vilas Bôas Fávero C, Haji-Akbari N, et al. The fractal aggregation of asphaltenes[J]. Langmuir, 2013, 29(28): 8799-8808.
[2]Fenistein D, Barre L. Experimental measurement of the mass distribution of petroleum asphaltene aggregates using ultracentrifugation and small-angle X-ray scattering[J]. Fuel, 2001, 80(2): 283-287.
[3] Barré, L., Simon, S., & Palermo, T. (2008). Solution properties of asphaltenes. Langmuir, 24(8), 3709-3717.




# Model: Parameters and Variables

**Number of each size clusters**

- Asphaltene system is power law distribution: $\mathbf{N} = \mathbf{A} \times M^{-\mathbf{a}}$ [1]
- The power law index(a) is calculated by using the average molecular weight of soluble asphaltenes. And normalize by solubility data.
- Power law index is from 0.2 to 1.66.

Molecular weight of soluble asphaltenes by SANX[2]

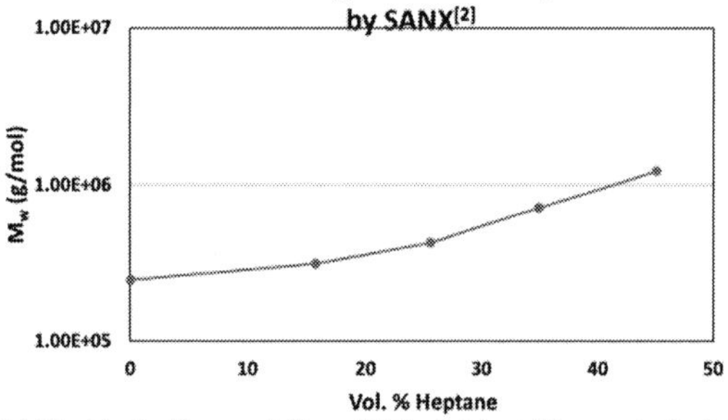


Asphaltenes molecular weight distribution [1]

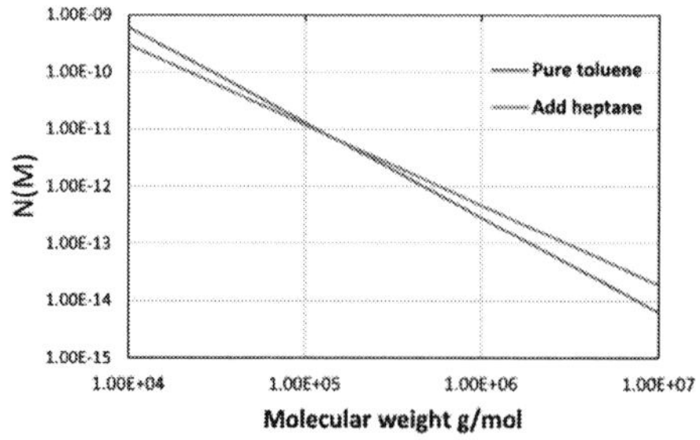



[1] Fenistein D, Barre L. Experimental measurement of the mass distribution of petroleum asphaltene aggregates using ultracentrifugation and small-angle X-ray scattering[J]. Fuel, 2001, 80(2): 283-287.
[2]Hoepfner M P, Vilas Bôas Fávero C, Haji-Akbari N, et al. The fractal aggregation of asphaltenes[J]. Langmuir, 2013, 29(28): 8799-8808.

# Simulation Results

- **Simulation produces similar trend as experimental results.**
- Insoluble asphaltene clusters shrink cause the intrinsic viscosity drop.
- When asphaltene clusters gain enough weight, shrink cannot lead the intrinsic viscosity to decrease.

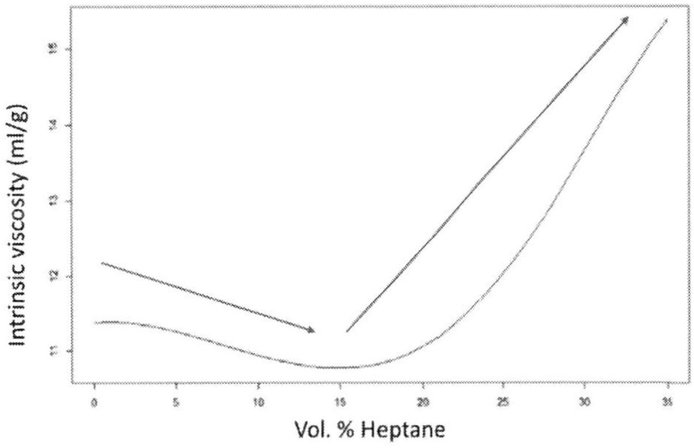


**Future Work:**

Collect all variables under same condition by experiments



# Experiment

## Goal

- **Expand our hypothesis by using different precipitant.**
- **Generate conditional results to verify our hypothesis and model.**

### Intrinsic viscosity measurements

Huggins equation $\eta_r = 1 + [\eta]\phi + k_H[\eta]^2\phi^2 + \cdots$

$$\eta_r = \frac{solution\ viscosity}{solvent\ viscosity}$$

$\phi$ Asphaltene vol fraction.

$k_H$ Huggins coefficient.

K Mark-Houwink coefficient

M Molecular weight

$[\eta]$ Intrinsic viscosity

Pure toluene

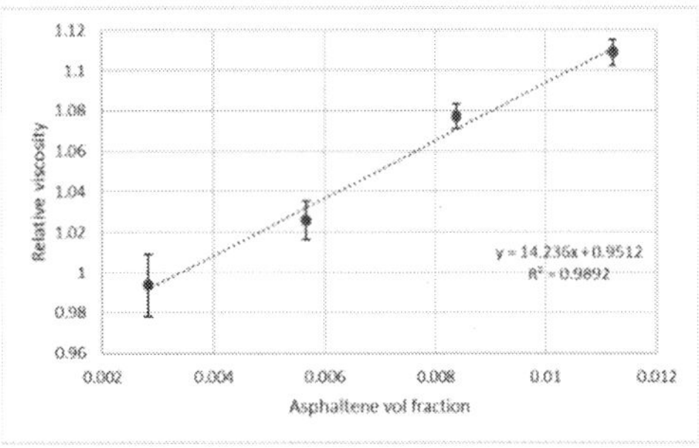


[1]Flory, Paul J. Principles of polymer chemistry. Cornell University Press, 1953.

# Experiment

## Materials

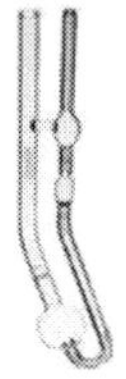


- Cannon capillary viscometer was used to measure viscosity, constant 0.002 cSt/s.
- Asphaltenes extracted from Athabasca Bitumen.
- **Solvent:** Toluene; **Precipitant:** n-Heptane and Decane; **Room temperature**
- Samples were centrifuged for large clusters removal.

**Experimental procedure**

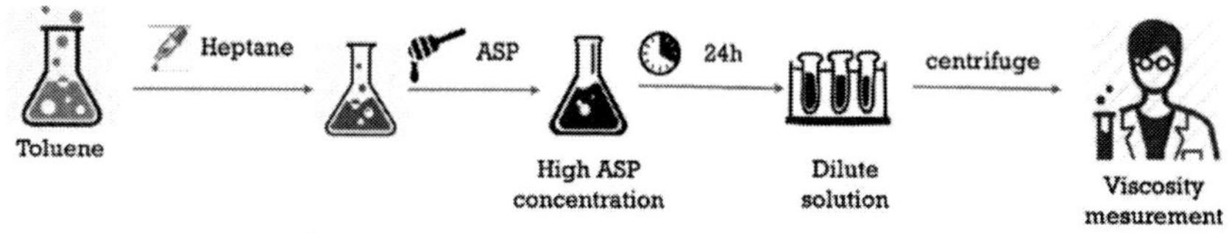


1. The capillary viscometer is the most accurate instrument for the ultra-low viscosity measurement.
2. Large clusters can possibly cause plugging problems.
3. Use mass balance to calibrate the asphaltenes concentration after centrifugation.



# Results

- Adding heptane into asphaltene toluene solution causes the intrinsic viscosity drop.
- When decane added into asphaltene toluene, the intrinsic viscosity is stable and the value is close to the value from asphaltenes in pure toluene.
- There are some explanations about the decane case.
  - Our measurement is not accurate.
  - Decane does not force enough soluble asphaltenes to precipitate.
  - Decane does not cause insoluble asphaltenes to shrink as much as heptane.
- Further study are under going.

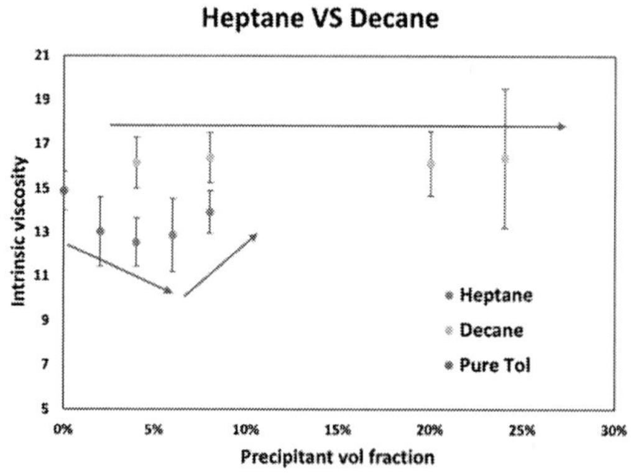

## Conclusion

Study the polydispersity effect on asphaltenes. Add precipitant into asphaltenes solution.

- The model produce the similar intrinsic viscosity trend as experimental data.
- Different precipitants may have different capacities on changing asphaltenes structure or solubility.

## Future Work

- Try to acquire more accurate asphaltenes solubility, molecular weight and fractal dimension from experiments or previous literature to modify our model.
- Use the fractionated asphaltenes to do experiments for investigating the behavior of different fractionated asphaltenes.



## Questions?

Weiyi.kong@utah.edu

THE UNIVERSITY OF UTAH

# 500034 Characterization of Gas Hydrate Slurry Viscosity Using a High Pressure Rheometer

**Abstract:**

**Characterization of Gas Hydrate Slurry Viscosity Using a High Pressure Rheometer**

*Ahmad AA Majid*[1,2], *David T Wu*[1,3], *Carolyn A Koh*[1*]

1. Center for Hydrate Research, Colorado School of Mines, Golden CO 80401, USA
2. Faculty of Chemical & Natural Resources Engineering, Universiti Malaysia Pahang, Malaysia
3. Department of Chemistry, Colorado School of Mines, Golden CO 80401, USA

## ABSTRACT

Gas hydrates are solid structures that contain small gas molecules (e.g. methane, ethane, propane, xenon) trapped inside molecular water cages. Hydrates typically form at high pressure and low temperature. As such, hydrates can form and plug oil and gas flowlines, and are therefore a major concern to the oil and gas industry. In this work, we performed in-situ gas hydrate slurry viscosity measurements using a high pressure rheometer. Measurements were performed using both a model emulsion and a crude oil emulsion as a function of water content, gas composition, subcooling (i.e., temperature and pressure) and shear rate. Measurements were conducted systematically in order to evaluate the effect of each variable on the viscosity of the system. Based on this work, we developed a model for relative viscosity as a function of the oil properties and hydrate volume fraction. Furthermore, initial investigations show that the viscosity profiles are significantly affected by the crystal structure (sI or sII) of the hydrate. It was also observed that the conversion of water to hydrate has a significant role on the agglomeration of hydrate particles in the system; whereby it is hypothesized that the presence of a free water layer may lead to significant agglomeration of hydrate particles. Comparisons between model and crude oil emulsion systems suggest that the natural surfactants in crude oil may help in preventing the agglomeration of hydrate particles.

# 500630 Avoiding Flowline Plugging: Emulsion+Ice+Wax+Hydrate, the Beast Flow Assurance Discipline Must Confront

**Abstract:**

*Solid generation and transportation in form of slurries has become one of the most important aspects of flow assurance engineering. Significant amount of work has been done in the past 40 years towards understanding and predicting the behavior of multiphase flow with solids; however, to date flow line plugging is still a challenge. There is no predictive tool available for use under all scenarios and there is still a lack of understanding of the interaction of the different flow assurance solids such as ice, wax and hydrates. To this list, emulsions should be counted as well since very high viscosity emulsions may stall flow.*

*This paper presents preliminary work done towards understanding system response with high viscosity emulsions. In addition to very high pressure losses, the systems are characterized by high shear heating effects. This means high temperature buildup in the areas of high shear such as close to the pipe wall. This phenomenon may be important in evaluating wax deposition since a negative radial temperature gradient is required. Four fluids were used to investigate this behavior and results showed that the temperature difference between the flowing fluid and the coolant in the annulus increased with fluid viscosity.*

*The second part of the paper presents pressure drop signature for hydrate transporting systems from fully dispersed to bedding onset to plugging. This section presents results from kerosene testing and examines factors affecting hydrate slurry transportation. Results showed that fully dispersed systems exhibit pressure drop curves running parallel to that of the carrier fluid. Onset on bedding was associated with pressure drop not changing with mixture velocity and occurred at a critical mixture velocity for a given hydrate volume fraction. Lastly, onset of plugging was characterized by increasing pressure drop with decreasing mixture velocity. For the low viscosity fluids, like kerosene, plugging results from solid fraction overburden in the system that the fluid cannot suspend.*

*The third part of the paper will discuss the effect of high viscosity emulsion on hydrate formation and transportability. Results showed that use of anti-agglomerants improves transportability of hydrates. For the high viscosity fluids, solids increase the slurry viscosity so high that the flowing mixture becomes mousse-like which eventually stalls the flow. This may be avoided by increasing the temperature of the flowing mixture.*

*Lastly, the paper will present preliminary work towards ice transportation, oil chemistry and hydrate morphology, and a review of the interactions of the different flow assurance solids.*

17AIChE
Annual Meeting, Minneapolis, MN

FOR RESEARCH CENTER HYDRATE

# Mass Transfer during Hydrate Formation in Water-Oil-Gas System

Dongdong Guo[1], Wenjia Ou[1], Fulong Ning[1,2*], Zhichao Liu[1], Wanjun Lu[1], **Ahmad A A Majid[2]** and Carolyn A. Koh[2]

1. China University of Geosciences, Wuhan, China
2. Colorado School of Mines, Golden, CO

October 31, 2017

## OUTLINE

- **Introduction**
- **Experimental**
- **Results and Discussion**
- **Conclusions**

# Introduction

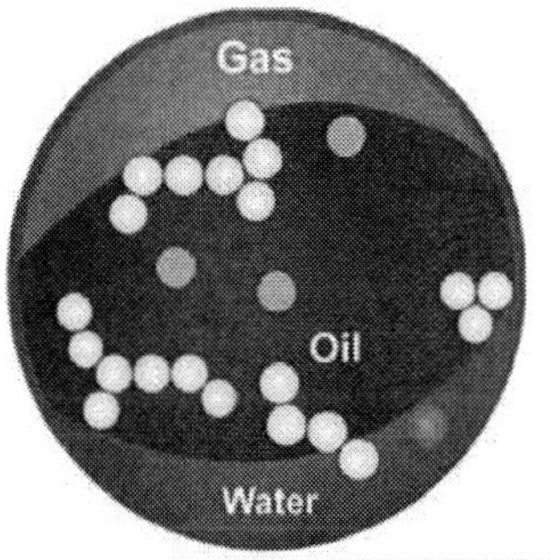


(Aman and Koh, 2016)

- Hydrate formation in the water-in-oil emulsion system concerns flow assurance and drilling safety(oil-based drilling fluid)
- However, the difference of hydrate formation between pure water and emulsion system, and the correspongding mass transfer mechanism in emulsion system are still unclear.

# Introduction

1 **Research content:** Mass transfer during hydrate formation in water-oil-gas system

2 **Research object:** understand the mass transfer during hydrate formation in water-oil-gas system;
investigate the effect of hydrate formation on the water-oil interface of water-oil-gas emulsion system

3 **Method:** experimental methods （In situ measurement technique based on Raman spectroscopy）.

# Experimental

## Experimental apparatus and materials

**Methane** (99.99%) used in this study was supplied by Air products;
**Water** was ultra-purified with a resistivity of 18.24 MΩ·cm.
**Mineral oil** （5#White Oil） Dongguan Lubricating Oil Co., Ltd.

JY/Horiba LabRam HR Raman system

# Experimental

## Experimental procedures

- Experimental conditions: **20MPa, 2℃**
- Sample preparation (in High Pressure Optic Cell, HPOC)

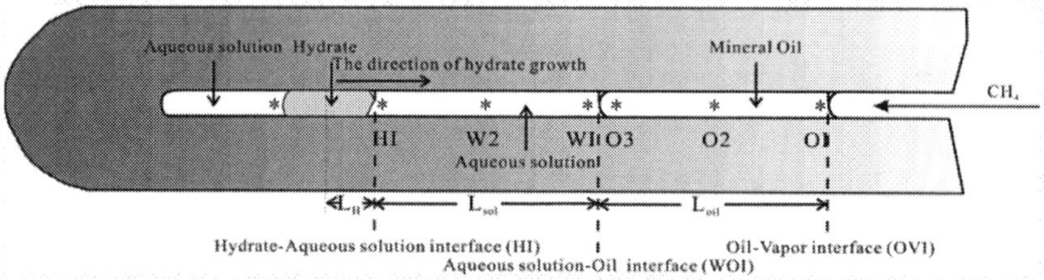


Hydrates were nucleated in the solution near the closed end of the optical tube .It is about 6 mm away from the W-O interface.

- In situ monitoring of the concentration distribution of dissolved methane

# Results And Discussion

## Hydrate formation and growth in Water-Oil System

- **Measured methane hydrate growth rates**

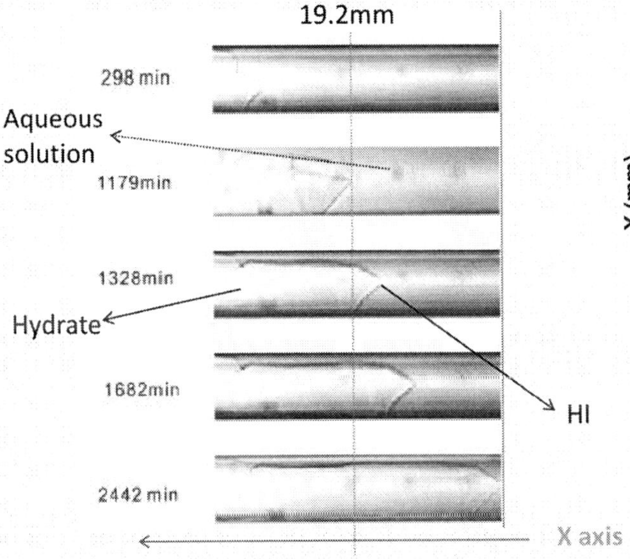


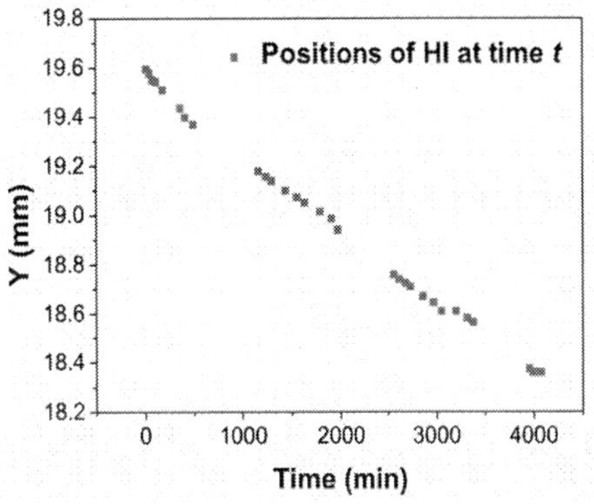


$$y_{HI} = -3.00717E-4t + 19.5497$$

# Results And Discussion

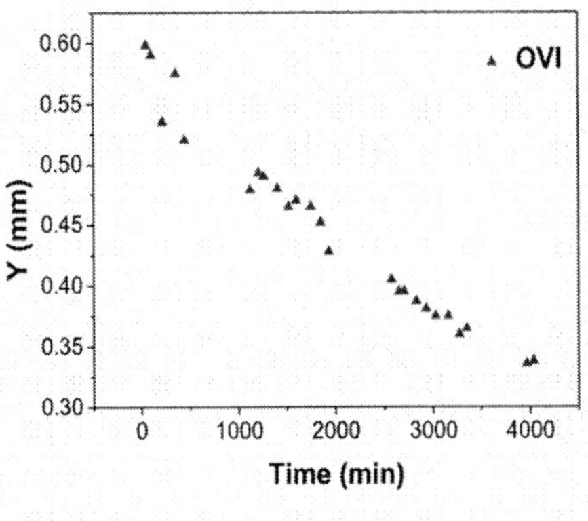


$$y_{OVI} = -6.29574E\text{-}5t + 0.57023$$

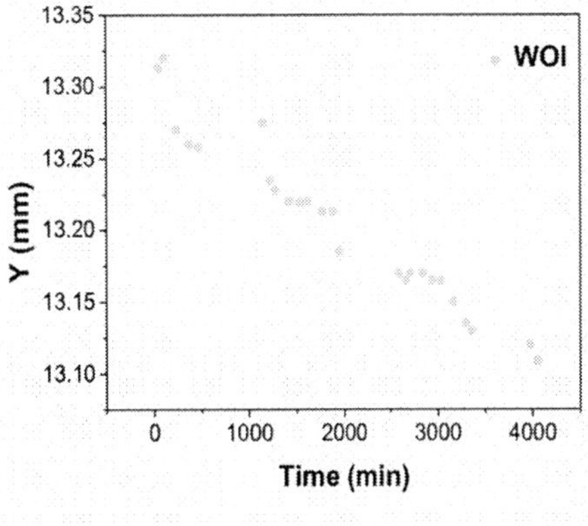


$$y_{WOI} = -4.66769E\text{-}5t + 13.29578$$

The position changes of O-V interface and W-O interface (moving towards the open end)

## Results And Discussion

Length of Aq. Solu. Section:

$$L_{sol,t} = y_{HI} - y_{WOI} = -2.5404E-4t + 6.25392$$

Length Increase of Hydrate Section:

$$L_{H,t} = 3.00717E-4t$$

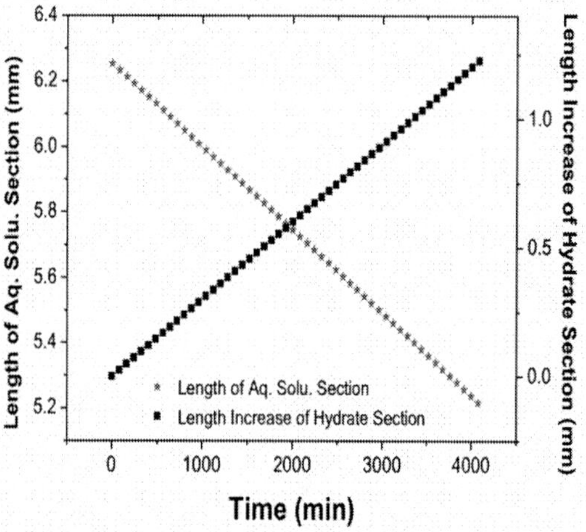



## Results And Discussion


### Measured methane concentration distributions

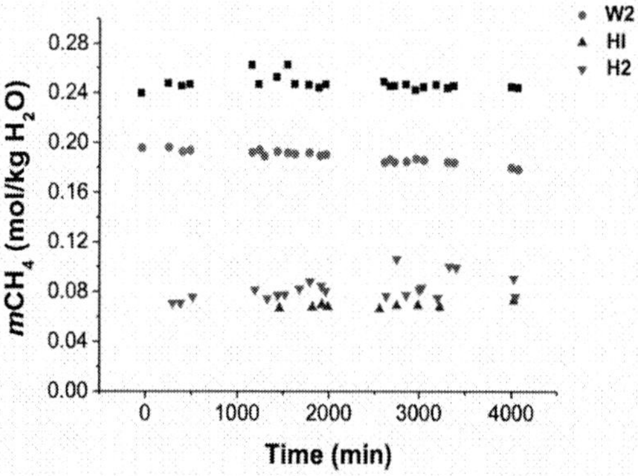


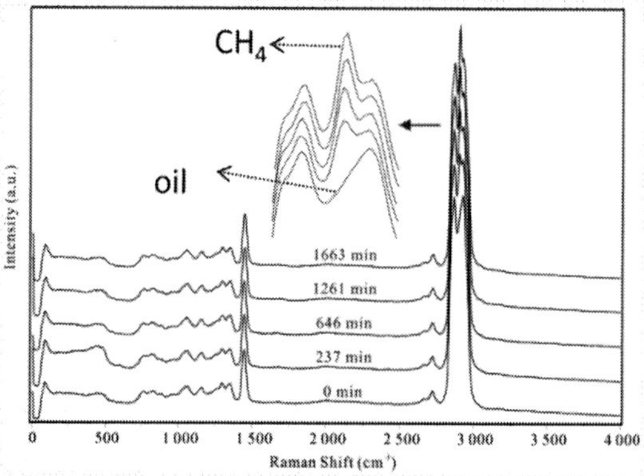


Concentrations of $CH_4$ at various sample spots of the aqueous solution and oil ($m_{CH4}$, mol/kg $H_2O$) during the growth of methane hydrate at 20MPa and 2℃.

# Results And Discussion

## Effect of oil on mass transfer

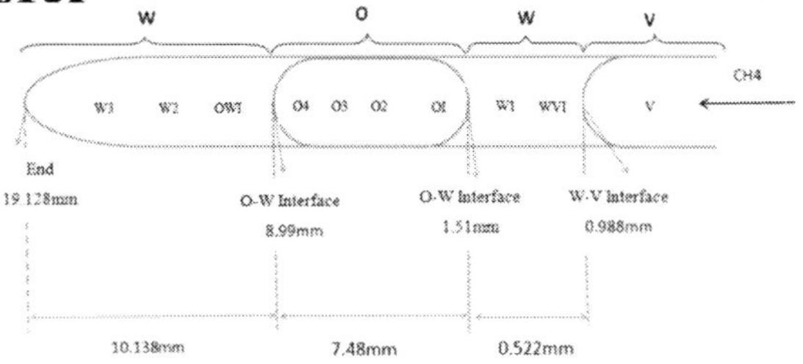


$D_{CH_4}$=0.7585e-09 $m^2/s$ in water

$D_{CH_4}$=0.924e-09 $m^2/s$ in oil

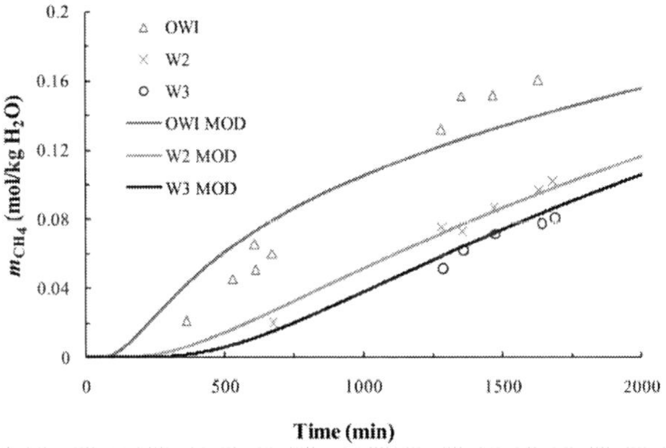


# Results And Discussion

## Methane flux to the crystal surface

The diffusion coefficients were calculated using the formula established by Guo et al:

$$D_{CH_4} = D_0\left(\frac{T}{T_S} - 1\right)^m$$

Thus the $CH_4$ flux through the cross-section (A) in a given time period can be expressed as follow:

$$\int JA\mathrm{d}t = -DA\rho_{sol}\Delta c\int\frac{1}{at+b}\mathrm{d}t = -DA\rho_{sol}\Delta c\int\frac{1}{(at+b)(at+b)'}(at+b)'\mathrm{d}t = -\frac{DA\rho_{sol}\Delta c}{a}\int\frac{1}{L_{sol,t}}\mathrm{d}L_{sol,t}$$

$$= -\frac{DA\rho_{sol}\Delta c}{a}(\ln L_{sol,t} + C) = -\frac{DA\rho_{sol}\Delta c}{a}(\ln(at+b) + C)$$

# Results And Discussion

**Kinetic process coupled surface reaction and mass transport**

the driving force can be expressed as:

$$\frac{\mathrm{d}n_H}{\mathrm{d}t} = K'A_{\mathrm{H}}\left(c - c_{\mathrm{eq}}\right) \qquad \frac{1}{K'} = \frac{1}{k_{\mathrm{d}}} + \frac{1}{k_{\mathrm{r}}}$$

$$\Pi_H = \frac{J_H}{J} = \frac{\mathrm{d}\left(L_{\mathrm{H,t}} / V_{\mathrm{m,H}} / \left(D\rho_{\mathrm{sol}}\Delta c / L_{\mathrm{sol},t}\right)\right)}{\mathrm{d}t}$$

**is equivalent to the growth rate of hydrate against the corresponding $CH_4$ diffusion flux. $\Pi_H$ are identical when the growth rates are constant .This shows that the growth rates are proportional to the diffusion fluxes of $CH_4$. The controlling process is considered to be diffusion, i.e. mass transfer, $k_d \ll k_r$ and $K' \approx k_d$ in this case.**

$$\frac{\mathrm{d}n_{\mathrm{H}}}{\mathrm{d}t} = \Pi_H JA \qquad K' \approx k_{\mathrm{d}} = \frac{\Pi_H DA}{L_{\mathrm{sol},t}A_{\mathrm{H}}}$$

# Results And Discussion

## The effect of hydrate formation on the water-oil interface

The change of water-oil interface during hydrate formation stimulated at different positions:

**The first excitation:** the excitation position is close to the end of the tube, and is about 6mm away from the W-O interface.

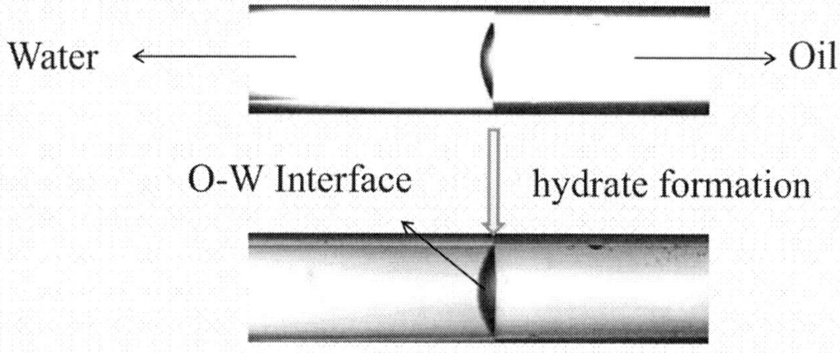


There is no obvious destruction of water-oil interface after hydrate formation

# Results And Discussion

**The second excitation :** the excitation position is about 1.3mm away from the W-O interface

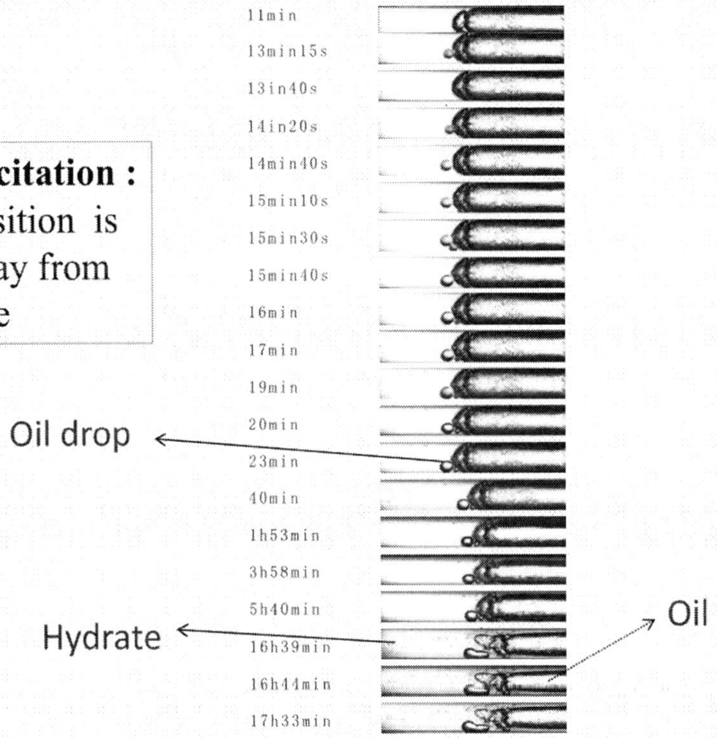


**The third time: the excitation position is close to the oil-water interface (about 0.8mm)**

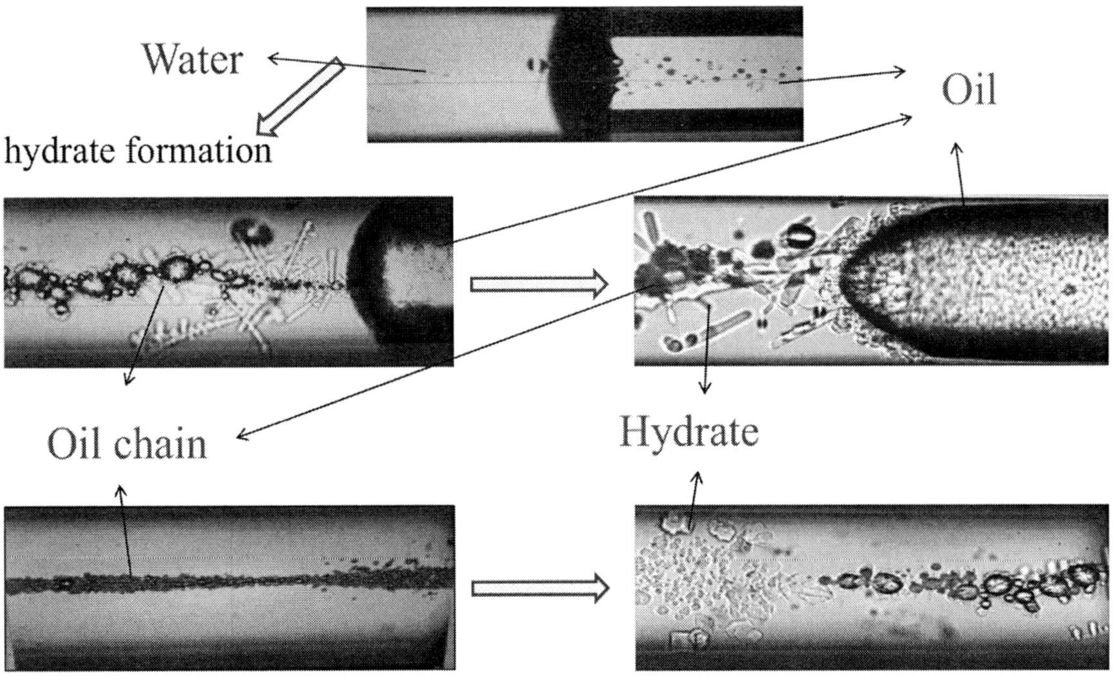

## Results And Discussion

### Comparison

The hydrate formation was stimulated at different locations of water part in the capillary tube. The results show that the oil near the water-oil interface enters into the water, even forms an oil chain in the water phase, and the original interface was destroyed.

It seems that the nearer the excitation position is to the oil-water interface, the greater the damage degree of the oil-water interface. This needs more experiments to verify. The surface of the oil droplet or the oil phase will form hydrate quickly and the oil will be wrapped up.

## Results And Discussion

### Discussion

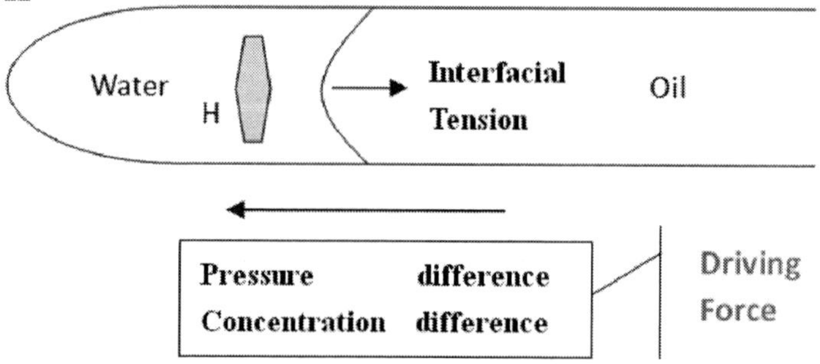


When hydrate is rapidly formed near the oil-water interface, a high concentration difference and pressure difference are formed between hydrate surface and oil-water interface.The driving force is greater than the additional pressure caused by interfacial tension, resulting in the destruction of oil-water interface.

$c_{WI}$=0.2476mol/kg $H_2O$, $c_{HI}$=0.083 mol/kg $H_2O$, $\Delta c$=0.1646 mol/kg $H_2O$
$\Delta P$=not yet determined

## Conclusions

- When the hydrate grows away from the oil-water interface, the growth rate of hydrate is controlled by diffusion, which is consistent with the growth law of hydrate in pure water
- Oil speeds up the diffusion rate of methane in oil- water-gas system.

## Conclusions

- When the hydrate formes near the oil-water interface, the oil-water interface will change or even destroy
- This means hydrate formation in the emulsion system may cause emulsion breaking, which would influence the stability of oil-based drilling fluid system
- This phenomena can be attributed to pressure difference and methane concentration difference under large drive force conditions

Gas Hydrate
A problem, a potential and an opportunity !
Thanks!
N S F C
National Natural Science
Foundation of China

# Enhancing Hydrate Inhibition Performance of Biomolecules by Doping with Synergents.

Tausif AlTamash[a] , Fahed Qureshi[a], Majeda Khraisheh[a1], Mohammad Ali Saad [a], Mert Atilhan [b]

[a] Department of Chemical Engineering, Qatar University, P.O.Box 2713, Doha, Qatar.

[b] Department of Chemical Engineering , Texas A&M University, College Station, TX, USA.

## Abstract:

The thermodynamic and kinetic inhibition effectiveness of the selected amino acids (AA): L-Alanine, Glycine, L-Histidine, L-Phenylalanine and L-Asparagine have been tested on the pure methane gas at different concentrations (1-5wt%) and pressure range of 40-120 bars using the rocking cell assembly with and without the presence of synergents /additive compounds such as polyethylene oxide (PEO) and Caprolactum (VCAP). The experimental results indicate that the selected AA perform better as a kinetic hydrate inhibitor (KHI) and the addition of PEO with these AA help to improve their KI effectiveness significantly.

**Key Words**: Amino Acids, Hydrates, Synergents, Methane Gas

## Introduction:

Natural gas is an economical source of fuel which is widely utilized for heating and providing power for domestic and industrial usage. It is also being utilized for generating electricity and its demand is increasing continuously all across the globe [1]. The natural gas is normally produce by drilling on the seabed and then transported from offshore to

[1] Corresponding author Tel.:+974 4403 4993; Fax: +974 4403 4131
Mobile: +974 30999747

onshore units via under water subsea lines under high pressure. As the temperature on the sea bed floor goes down and the pressure within the subsea line is high, the formation of natural gas hydrates (NGH) takes place which can lead to unwanted bloackages [2, 3].

NGH are crystalline compounds that are formed under higher pressure and low temperature conditions when small gas molecules like methane and ethane get trapped within the water molecules and form clusters. The X-ray analysis [4, 5] show that gas hydrate are typically found in three structures such as cubic structure I (sI), cubic structure II (sII) and hexagonal structure H (sH). The small gas molecules such as $CH_4$, and $CO_2$ form sI cubic structure, the $C_3H_8$ and iso-$C_4H_{10}$ for sII cubic structure and larger molecules form hexagonal structure (sH) [6].

The use of chemical inhibitors is considered to be one of the most prominent ways to prevent the hydrate formation in the subsea lines. These chemical inhibitors are classified as thermodynamic hydrate inhibitors (THI), kinetic hydrate inhibitors (KHI) and Anti-agglomerants (AA). The THI prevent the hydrate formation by shifting hydrate liquid vapor equilibrium (HLVE) curve to lower temperatures[7] and the KHI function by delaying the hydrate formation time [8]. AA does not inhibit hydrate formation but they prevent the growth of hydrate crystals [9]. The THI like methanol are widely used in industry but they required in bulk quantity (> 30 wt %) and they cannot be easily disposed of into the environment [10]. Therefore, the research interest has shifted towards the search of inhibitors that are environmentally friendly and required in low dosage.

In the last years, the biodegradable chemical compounds like urea [11], chitosan [12], proteins [8] and synthetic bio-molecules have been tested as the gas hydrate inhibitors [13-16]. Recently, amino acids (AA) have gain attention to be as a hydrate inhibitors as they are known to be environmentally friendly, biodegradable and cheap in comparison to ionic liquids (ILs) [15].

Amino Acids (AA) are used for protein synthesis and are considered to be the building blocks of life [17]. AA are increasingly being used as an animal feed, flavoring agents, cosmetics ingredients, sports supplements and other pharmaceutical drugs [18]. They are considered to have negligible impact on the aquatic life in sea which makes it easier to dispose them off easily [19-21]. The AA act as a zwitterion and contain carboxylic (-COOH) and ammonium (-$NH_2$) groups attached to the center carbon atom. Some of the addition groups such as amide, phenyl and imidazolium are also attached with AA that helps in the classification of AA as acidic, basic or neutral. In addition to that the AA are also classified as hydrophilic or hydrophobic based on the attached group and their physical chemical properties. The AA can disrupt the hydrogen bonding in the water molecules via electrostatic force of attraction. According to Raman Spectroscopy analysis the it's the hydrophilic or hydrophobic moiety of AA that strengthen or disrupt the hydrogen bonding in water cages [22]. It has also been deduced that it's the electrostatic interactions between inhibitor and water molecules that lowers the activity coefficient of the water and leads to thermodynamic inhibition [23].

So the objective of this work is to test the thermodynamic and kinetic inhibition effectiveness of the selected amino acids (AA): L-Alanine, Glycine, L-Histidine, L-Phenylalanine and L-Asparagine on the pure methane gas at different concentrations (1-5wt%) and pressure range of 40-120 bars using the rocking cell assembly with and without the presence of synergents /additive compounds such as polyethylene oxide (PEO) and Caprolactum (VCAP). The most part of this work has already been published in the relevant ACS journal.[24]

## Experimental Section:

### Reagents

The pure methane gas (99.9 % purity) used in this work was acquired from Buzware Company, Doha (Qatar). The amino acids and synergent compounds used in this work were acquired from Sigma Aldrich. The list of amino acids and synergent compounds used in this work are shown in Table 1. All the sample solutions were prepared using the distilled water.

### Equipment and Procedure

All the experiments were conducted using a Rocking cell assembly (RC-5) of five cells purchased from PSL SystemtechniK GmbH, Germany. RC-5 can be operated up to 200 bars (2900 psi) and within temperature zone of −10 °C to 60 °C. RC-5 consists of 5 stainless steel cells attached to a rig submerged in the water bath. Each cell has a volume of 40.13 $cm^3$ and contains a stainless steel (ø = 17 mm), which moves back and forth to provide necessary agitation [25]. A sample solution of 15 ml is added inot each cell and then the cells are pressurized from 40-120 bars sequentially. Once the cells are pressurized the RC-5 is stabilized at 20 °C for 1 hr and then the cooling process is initiated where the cells are cooled from 20 °C to 2 °C in period of 9 hrs, followed by an isothermal step of 24 hrs where temperature is kept constant at 2 °C and finally the cells are heated slowly from 2 °C to 20 °C to dissociate the hydrate crystals. The Density of some aqueous solution were also measured at different temperatures using Anton Paar DMA 4500M. The density and temperature accuracy of the device was reported to be as ±0.00005 g $m^{-3}$ and ±0.05 K respectively[26]. The detail experimental procedure and the extraction procedure for the hydrate dissociation and hydrate formation points can be found in the previously published works [24, 27, 28].

Table 1: List of AA and Synergents used in this work

| *Amino Acid* | *Structure* | *M.W (g/mol)* | *Purity (%)* | *Supplier* |
|---|---|---|---|---|
| Glycine | 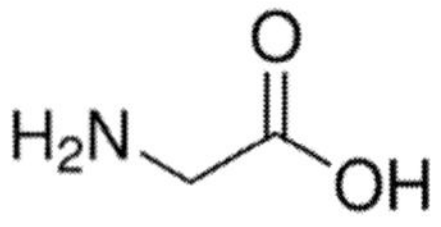 | 75.06 | ≥98.5 | Sigma-Aldrich |
| L-Alanine | 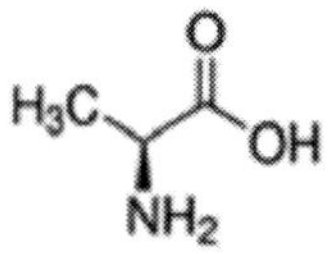 | 89.09 | ≥99.5 | Sigma |
| L-Histidine | 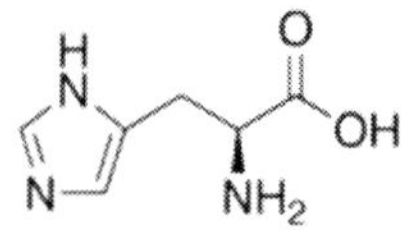 | 155.15 | ≥99.5 | Sigma |
| L-Asparagine | 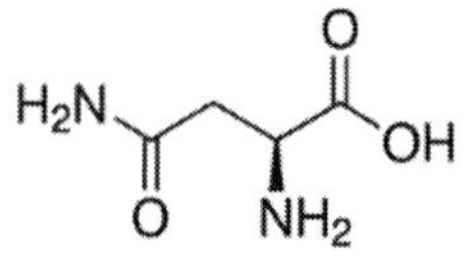 | 132.12 | ≥98.0 | Sigma |
| L-Phenylalanine | 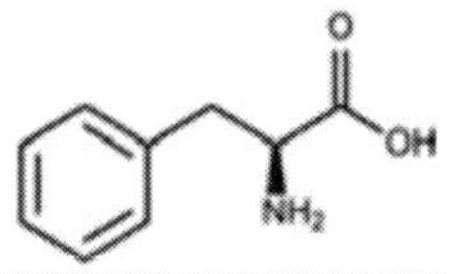 | 165.19 | ≥99.0 | Sigma |

| | | | | |
|---|---|---|---|---|
| Polyethylene Oxide | 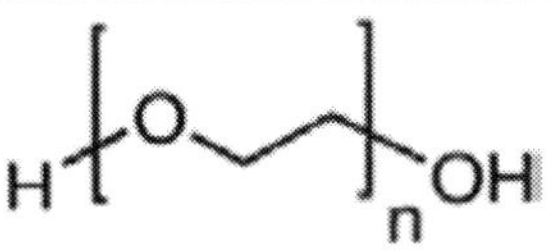 | ≈ 100,000 | ≥98.0 | Sigma-Aldrich |
| VinylCaprolactum | 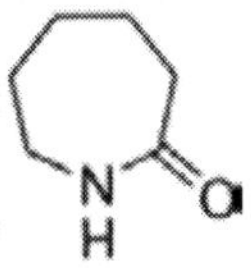 | 139.19 | ≥98.0 | Sigma-Aldrich |

## Results and Discussion:

### Thermodynamic inhibition (TI) of $CH_4$ Hydrate via AA:

Initially the low dosage experiments were conducted using 1 wt% of each amino acids (AA). It was found that no TI takes at the low dosage of AA which gives a slight indication that the low dosage AA were incapable of inhibiting $CH_4$ hydrates. Therefore, based on their solubility in water the AA Glycine, L-Alanine and L-Histidine were tested at higher concentration (4-5wt%) .

At the higher concentration the selected AA such as 5wt% Glycine and 5wt% L-Alanine showed observable TI effect. The 5wt% L-Alanine provided the hydrate suppression temperature within the range of 0.68 ± 0.2 °C to 1.1± 0.2 °C (38-110 bars) and the 5wt% Glycine provided the hydrate suppression temperature within the range of 0.62 ± 0.2 °C to 0.79 ± 0.2 °C (39-96 bars). The 4wt% Histidine was found to be the least effective with the hydrate suppression temperature of 0.3 ± 0.2 °C (~114 bars). So in terms of effectiveness as the THI at higher concentration the selected AA can be listed as [Figure 1]:

**5wt% L-Alanine > 5wt% Glycine > 4wt% Histidine**

The density/viscosity measurements show that at higher concentration of AA the availability of inhibitor molecules per unit volume of water increases. So the rate of hydrogen bonds disruption increases and inhibition effectiveness of the AA increases accordingly. J-H. Sa and coworkers [16] investigated the TI effectiveness of amino acids on $CO_2$ hydrate formation and found that inhibition effectiveness of AA increases with the concentration of AA. They also found that the AA with lower solubility in aqueous solution provides no inhibition effect and they also reported Glycine and L-Alanine as a good candidate for thermodynamic hydrate inhibition.

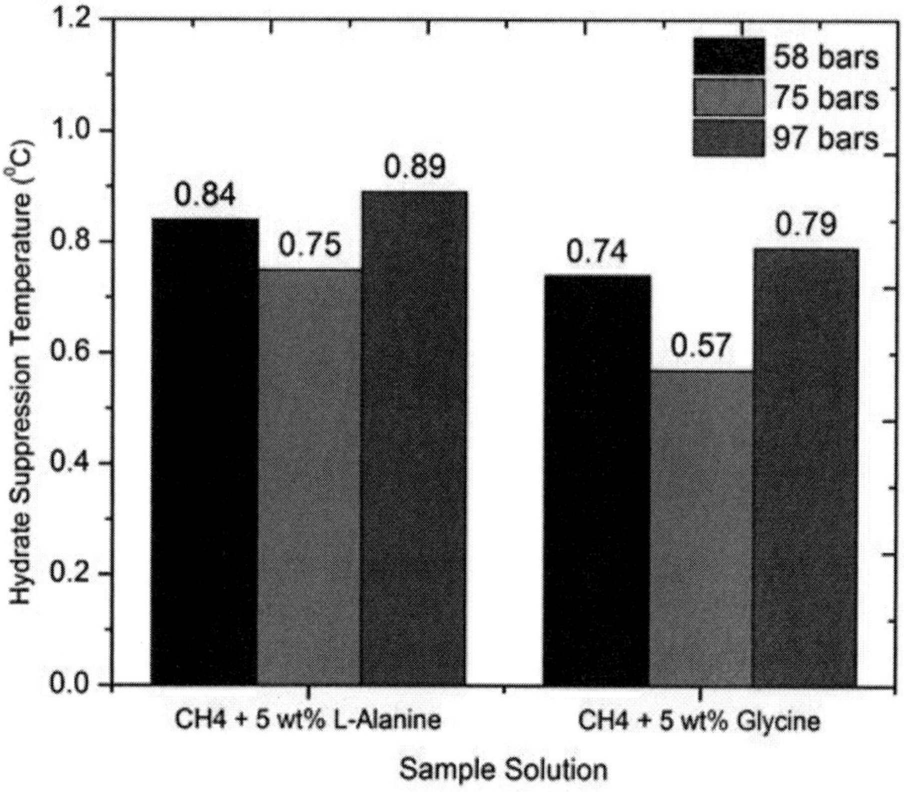


Figure 1: Shows the methane ($CH_4$) hydrate suppression temperature in the presence of amino acids (AA) 5wt% L-Alanine and 5wt% Glycine at different pressures with error of $\pm$ 0.2 $^{\circ}$C. The figure is adapted from the published work.[24]

## Kinetic inhibition (KI) for $CH_4$ Hydrate via AA:

The selected AA showed no TI effect low concentration of 1wt%. The graphical illustrations were plotted with time on the y-axis and the pressure on the x-axis. The illustrations clearly show that most AA provided observable KI effect at the low concentration of 1wt%, which gives a slight indication that the selected AA are more suited as a KHI. At the 1wt% of AA at 40 bars, the phenyl alanine was found to be the most effective KHI which delayed the hydrate induction time from 9.2 to 10 hrs and the Histidine was found to be the least effective KHI showed no significant KI effect. So in terms of effectiveness as KHI at low concentrations the AA can be listed as:

**L-Phenylalanine > L-Asparagine ≈ Glycine ≈ L-Alanine > L-Histidine**

L-Phenylalanine provided maximum delay in hydrate induction time among other inhibitors at low concentration of 1wt%.

The further experiments were conducted using 5wt% of Glycine, 5wt% L-Alanine and 4wt% Histidine. At the higher concentration and low pressure (~40 bars) the L-Alanine was found to be the most effective KHI by delaying hydrate induction time from 9.2 to 10.1 hr and the L-Histidine was again found to be the least effective KHI. So in terms of effectiveness as KHI at high concentrations the AA can be listed as:

**L-Alanine > Glycine > L-Histidine**

Sa et al. [15, 22] also investigated the KI effect of AA on the hydrate formation and found that Glycine and L-Alanine were a strong candidate for as a KHI. They also found no significant increase delay in hydrate induction time on increasing the concentration of AA. Also on increasing the concentration of amino acids they did not found significant change in nucleation. Roosta et.al [14] and Naeji [13] also found Glycine as an effective KHI for the $CO_2$ and tetrahydrofuran hydrates.

### Enhancement in inhibition effectiveness of AA using synergents

In order to enhance the KI and TI effect of the AA the synergents compounds Caprolactum (VCAP) and polyethylene oxide (PEO) were added with the selected AA. The 1wt% of VCAP and PEO were added with 5 wt % of Glycine, 5 wt.% of L-Alanine and 4 wt.% of L-Histidine at different pressures using same experimental protocol. The VCAP show a slight TI effect, but the PEO had no effect on the improvement of TI performance of AA.

However, PEO showed a significant improvement in the KI performance of the selected AA. At lower pressures (~55 bars), the 1wt% PEO + 5wt% L-Alanine mixture delayed the hydrate induction time from 7.3± 0.5 to 35± 0.5 hrs. Similarly, the mixtures 1wt% PEO + 5wt% Glycine and 1wt% PEO + 5wt% L-Histidine delayed the hydrate induction from 7.3± 0.5 to 30± 0.5 hr and 7.3± 0.5 to 27± 0.5 hrs respectively [Figure 2].

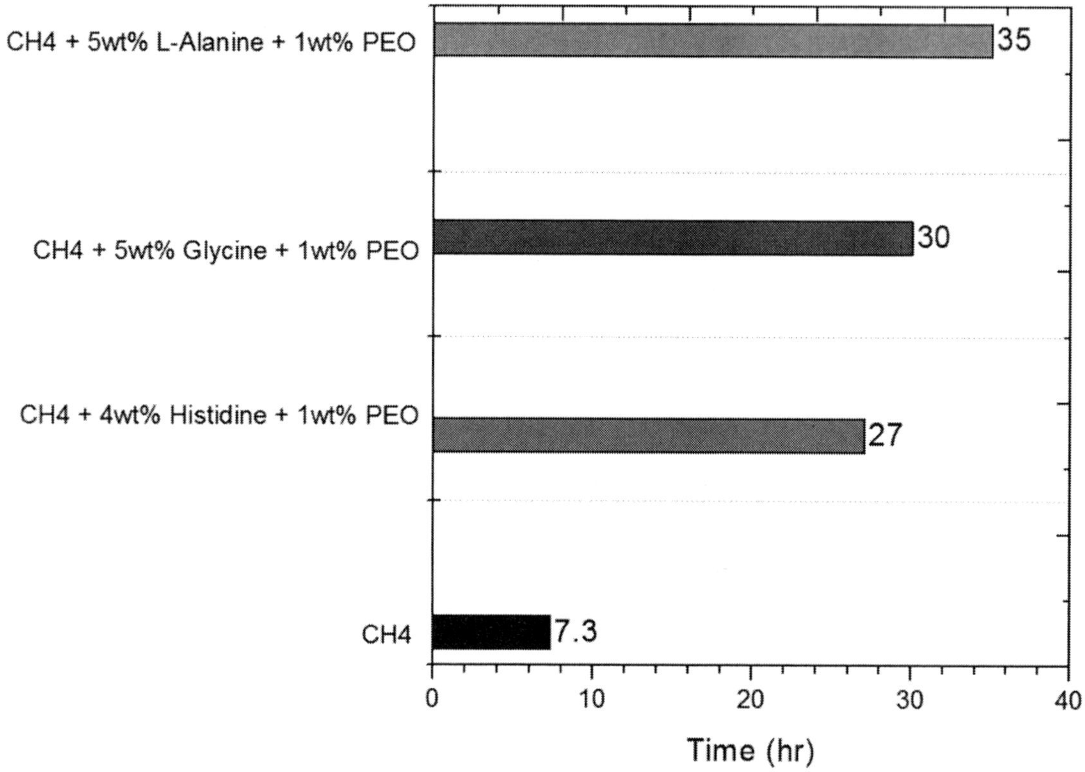


**Figure 2: The kinetic inhibition (KI) of methane (CH4) hydrates in the presence of amino acids (AA) and synergent PEO at the pressure of 55 bars with error ± 0.5 hr.**

The water soluble PEO consists of a hydrophilic oxygen atoms at the center and the polymerized hydrophobic ethylene chain connected to the oxygen atom. The hydrophobic polymeric chain of PEO repels the water molecules away from $CH_4$ via van der waal hydrophobic interactions and acts as a shield for the water molecules. The other side of PEO consists of an hydrophilic oxygen atoms that forms hydrogen bonding with water molecules via electrostatic force of attraction causing delay in the hydrate formation. Kang et al. [29] also provided the similar explanation for the inhibition mechanism for the polymer PVCAP and reported twice a delay in hydrate induction time of $CH_4$ with addition of PVCap with Ionic liquid [HEMP] [$BF_4$].

## Conclusion:

The thermodynamic inhibition (TI) and kinetic inhibition (KI) effectiveness of the selected amino acids (AA): L-Alanine, Glycine, L-Histidine, L-Phenylalanine and L-Asparagine have been tested on the pure methane gas at different concentrations (1-

5wt%) and pressure range of 40-120 bars using the rocking cell assembly with and without the presence of synergents /additive compounds such as polyethylene oxide (PEO) and Caprolactum (VCAP). The experimental results indicate that the selected AA can act as both thermodynamic hydrate inhibitor (THI) and kinetic hydrate inhibitor (KHI). However, the AA are more suited to act as KHI at higher concentration. It was also found that the addition of low dosage of PEO can significantly improve the KHI perform better as a kinetic hydrate inhibitor (KHI) and the addition of PEO with these AA help to improve their KI effectiveness significantly. The Glycine and L-Alanine were found to be the most effective KHI among the selected AA. The part of this work has already been published in the relevant journal.[24]

## Acknowledgement:

Author Fahed Qureshi would like to acknowledge GSRA. This work was made possible by GSRA # 2-1-0603-14012 from the Qatar National Research Fund (a member of Qatar Foundation). The statements made herein are solely the responsibility of the authors

## References:

1. Brown, A.S., et al., *Traceable Reference Gas Mixtures for Sulfur-Free Natural Gas Odorants.* Analytical Chemistry, 2014. **86**(13): p. 6695-6702.
2. Kvamme, B., et al., *Hydrate Formation during Transport of Natural Gas Containing Water and Impurities.* Journal of Chemical & Engineering Data, 2016. **61**(2): p. 936-949.
3. Aman, Z.M. and C.A. Koh, *Interfacial phenomena in gas hydrate systems.* Chemical Society Reviews, 2016. **45**(6): p. 1678-1690.
4. Loveday, J.S., et al., *Stable methane hydrate above 2[thinsp]GPa and the source of Titan's atmospheric methane.* Nature, 2001. **410**(6829): p. 661-663.
5. Koh, C.A., *Towards a fundamental understanding of natural gas hydrates.* Chemical Society Reviews, 2002. **31**(3): p. 157-167.
6. Sloan, E.D., *Fundamental principles and applications of natural gas hydrates.* Nature, 2003. **426**(6964): p. 353-363.
7. Jager, M.D., C.J. Peters, and E.D. Sloan, *Experimental determination of methane hydrate stability in methanol and electrolyte solutions.* Fluid Phase Equilibria, 2002. **193**(1–2): p. 17-28.
8. Kelland, M.A., *History of the Development of Low Dosage Hydrate Inhibitors.* Energy & Fuels, 2006. **20**(3): p. 825-847.
9. Kumar, A., T. Sakpal, and R. Kumar, *Influence of Low-Dosage Hydrate Inhibitors on Methane Clathrate Hydrate Formation and Dissociation Kinetics.* Energy Technology, 2015. **3**(7): p. 717-725.

10. Xiao, C., N. Wibisono, and H. Adidharma, *Dialkylimidazolium halide ionic liquids as dual function inhibitors for methane hydrate.* Chemical Engineering Science, 2010. **65**(10): p. 3080-3087.
11. Muromachi, S., et al., *Phase equilibrium for clathrate hydrate formed in methane + water + urea system.* Fluid Phase Equilibria, 2015. **398**: p. 1-4.
12. Xu, Y., M. Yang, and X. Yang, *Chitosan as green kinetic inhibitors for gas hydrate formation.* Journal of Natural Gas Chemistry, 2010. **19**(4): p. 431-435.
13. Naeiji, P., A. Arjomandi, and F. Varaminian, *Amino acids as kinetic inhibitors for tetrahydrofuran hydrate formation: Experimental study and kinetic modeling.* Journal of Natural Gas Science and Engineering, 2014. **21**: p. 64-70.
14. Roosta, H., et al., *Inhibition properties of new amino acids for prevention of hydrate formation in carbon dioxide–water system: Experimental and modeling investigations.* Journal of Molecular Liquids, 2016. **215**: p. 656-663.
15. Sa, J.-H., et al., *Hydrophobic amino acids as a new class of kinetic inhibitors for gas hydrate formation.* Scientific Reports, 2013. **3**: p. 2428.
16. Sa, J.-H., et al., *Amino Acids as Natural Inhibitors for Hydrate Formation in CO2 Sequestration.* Environmental Science & Technology, 2011. **45**(13): p. 5885-5891.
17. Riyazuddeen, T. Altamash, and A. Coronas, *Interactions in l-histidine/l-glutamic acid/l-tryptophan/glycylglycine + 2 mol L−1 aqueous KCl/KNO3 systems at different temperatures: An isothermal compressibility study.* Thermochimica Acta, 2012. **543**: p. 313-317.
18. Ivanov, K., et al., *Biotechnology in the Production of Pharmaceutical Industry Ingredients: Amino Acids.* Biotechnology & Biotechnological Equipment, 2013. **27**(2): p. 3620-3626.
19. Barouni, K., et al., *Some amino acids as corrosion inhibitors for copper in nitric acid solution.* Materials Letters, 2008. **62**(19): p. 3325-3327.
20. Hluchan, V., B.L. Wheeler, and N. Hackerman, *Amino acids as corrosion inhibitors in hydrochloric acid solutions.* Materials and Corrosion, 1988. **39**(11): p. 512-517.
21. Dehdab, M., M. Shahraki, and S.M. Habibi-Khorassani, *Theoretical study of inhibition efficiencies of some amino acids on corrosion of carbon steel in acidic media: green corrosion inhibitors.* Amino Acids, 2016. **48**(1): p. 291-306.
22. Sa, J.-H., et al., *Gas hydrate inhibition by perturbation of liquid water structure.* Scientific Reports, 2015. **5**: p. 11526.
23. *Hydrate Formation and Dissociation Processes*, in *Clathrate Hydrates of Natural Gases, Third Edition*. 2007, CRC Press. p. 113-187.
24. Altamash, T., et al., *Gas hydrates inhibition via combined biomolecules and synergistic materials at wide process conditions.* Journal of Natural Gas Science and Engineering, 2017.
25. Qureshi, M.F., et al., *Gas Hydrate Prevention and Flow Assurance by using Mixtures of Ionic Liquids and Synergent Compounds: Combined Kinetics and Thermodynamic Approach.* Energy & Fuels, 2016.
26. Ullah, R., et al., *A detailed study of cholinium chloride and levulinic acid deep eutectic solvent system for CO2 capture via experimental and molecular simulation approaches.* Physical Chemistry Chemical Physics, 2015. **17**(32): p. 20941-20960.
27. Qureshi, M.F., et al., *Gas Hydrate Prevention and Flow Assurance by Using Mixtures of Ionic Liquids and Synergent Compounds: Combined Kinetics and Thermodynamic Approach.* Energy & Fuels, 2016. **30**(4): p. 3541-3548.

28. Qureshi, M.F., et al., *High-pressure gas hydrate autoclave hydraulic experiments and scale-up modeling on the effect of stirring RPM effect.* Journal of Natural Gas Science and Engineering, 2017. **38**: p. 50-58.
29. Kang, S.-P., et al., *Unusual synergy effect on methane hydrate inhibition when ionic liquid meets polymer.* RSC Advances, 2013. **3**(43): p. 19920-19923.

Session: Fundamentals and Applications of Flow Assurance

# Investigation on the effect of wax on hydrate formation characteristics and flow properties in a high pressure flowloop

Yang Liu, Bohui Shi*, Ye Zhang, Lin Ding, Yu Yong, Shangfei Song, Wei Wang, Jing Gong*

*Email: sbh228@126.com; ydgj@cup.edu.cn

CHINA UNIVERSITY of PETROLEUM, BEIJING

2017 AIChE ANNUAL MEETING

AIChE The Global Home of Chemical Engineers



CHINA UNIVERSITY of PETROLEUM, BEIJING

## 2017 AIChE ANNUAL MEETING

AIChE
The Global Home of Chemical Engineers

### Backgrounds

- **Deep-water flow assurance**
  - wax deposition
  - hydrate plugging
  
  …
- **Coexistence state**
  - emergent shut-in
  - throttle regulation of low oil-gas ratio products
  
  …
- **Independent research**
  
  → **Decoupling research**

Complicated!

Wax deposition

Hydrate plugs

CHINA UNIVERSITY of PETROLEUM, BEIJING



## 2017 AIChE ANNUAL MEETING

AIChE
The Global Home of Chemical Engineers

### Experimental methods

- **High pressure flowloop**
  - Details L: 30 m
    D: 1 inch
    Tem.: -20℃~100 ℃
    Pre.: 0~15MPa
    Flow rate: 0~1900kg/h
  - Facilities
    
    Sensors (P, T, Q…)
    FBRM
    PVM
    High pressure visual window
    …

➡ Parametric & Visual information

Micro & Macro information

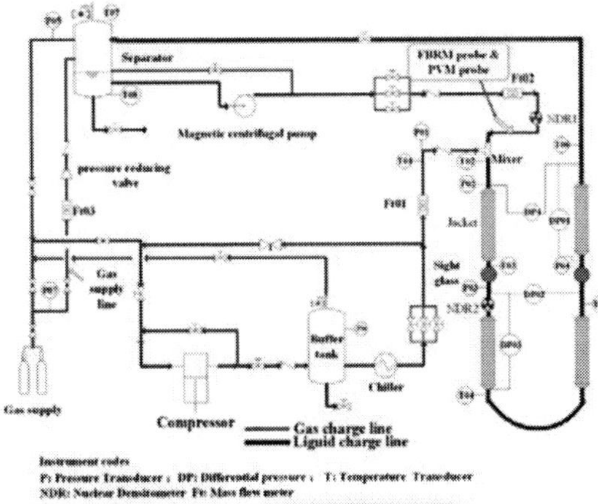


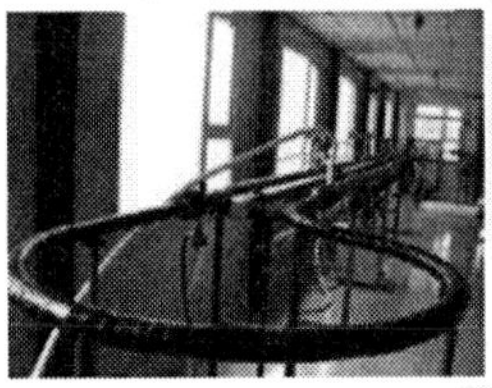

FBRM

PVM

CHINA UNIVERSITY of PETROLEUM, BEIJING

## Experimental methods

- **Materials**
  - 80vol%diesel oil + 20vol%water + 1wt%AA + 0.75wt%wax + 50bar
- **Measurement of properties**
  - WAT by DSC

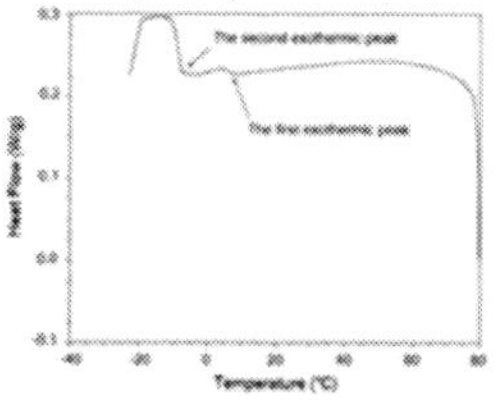


  - Viscosity by Rheometer
  - Composition by HPGC
- **Procedures**

DSC Rheometer HPGC

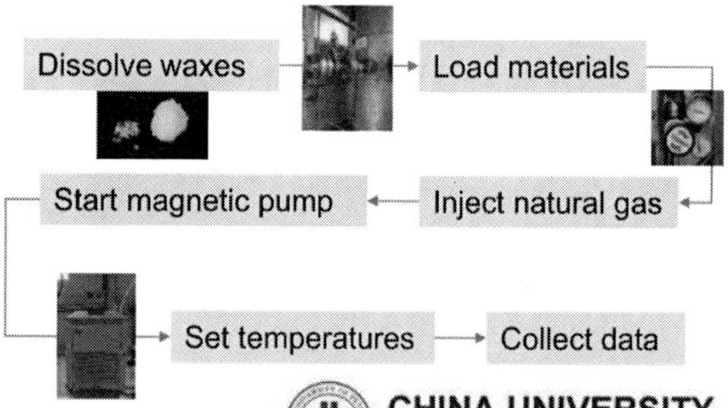






## Results and discussion

- **Effects of wax on hydrate induction time**
  - Wax precipitated before hydrate formation
  - Definition of induction time

| Wax content | Water bath temperature(°C) | Induction time(h) | Subcooling degree*(K) |
|---|---|---|---|
| without wax | -1 | 0.70±0.10 | 5.80±0.10 |
| | 1 | 1.42±0.10 | 5.60±0.10 |
| 0.75% | -1 | 1.15±0.10 | 6.94±0.10 |
| | 1 | 1.50±0.10 | 6.62±0.10 |
| | 3 | 3.15±0.10 | 6.23±0.10 |
| | 5 | 4.65±0.10 | 5.74±0.10 |
| 1.00% | -3 | 5.54±0.10 | 8.40±0.1 |
| 2.89% | 0 | -- | -- |

  - Wax crystals prolonged induction times

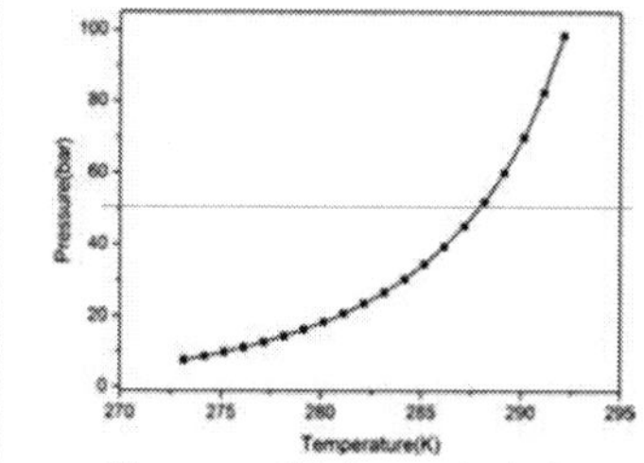


Phase equilibrium calculation of hydrate formation

0.75%&1%: Hydrate equilibrium tem. > WAT >hydrate formation tem.

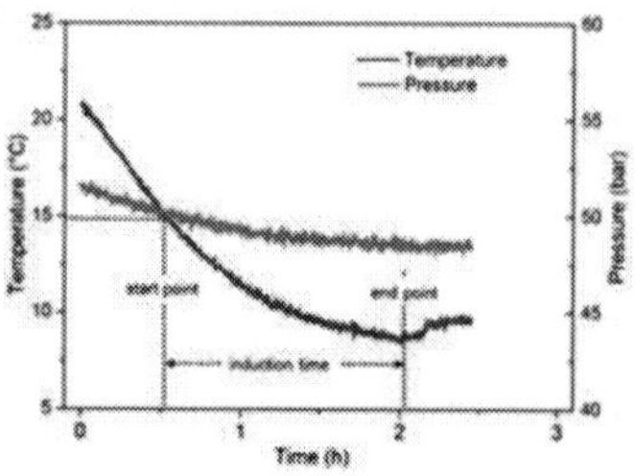


Induction time of 0.75% under 1 °C set temperature

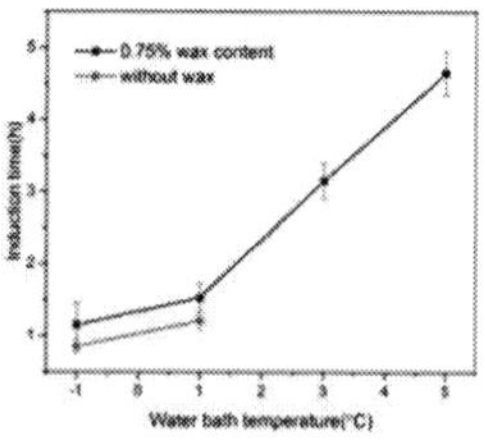


Induction times of 0.75% and non-wax

## Results and discussion

- **Effects of wax on hydrate induction time— Mechanisms**

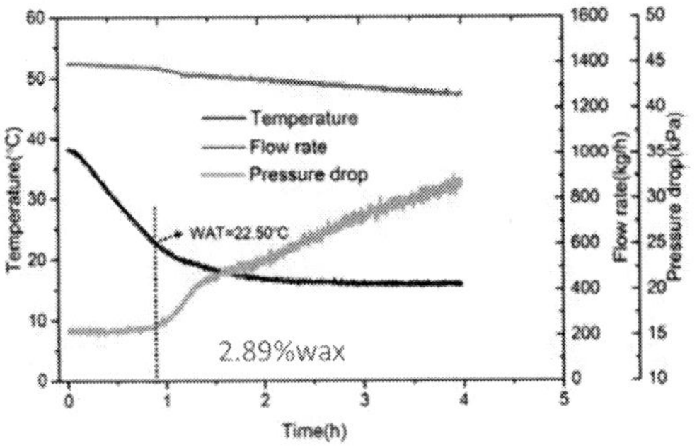


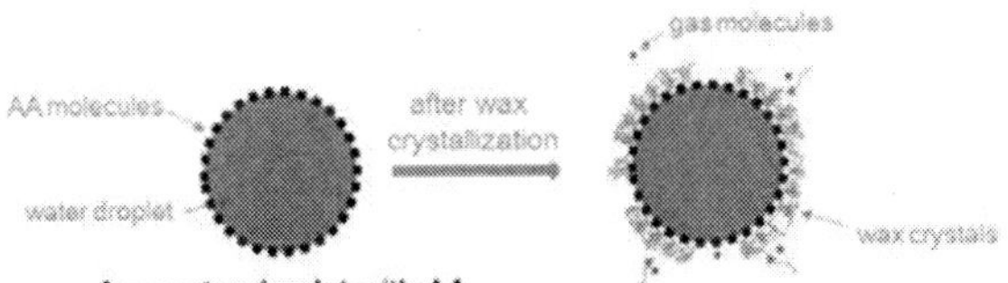


A: a water droplet with AA molecules absorbed

B: a water droplet with AA molecules and wax crystals absorbed

- Wax deposition layer → reduce the cooling efficiency of water baths
- Wax crystals absorbed → reduce the surface area for nucleation



2017 AIChE ANNUAL MEETING
AIChE
The Global Home of Chemical Engineers

## Results and discussion

- **Effects of wax on flow properties**
  - 3 stages vs. 2 stages
  - Definition of pipeline plugging
  - Low hydrate volume fraction while pipeline plugging
  - Sudden jump of pressure drop and flow rate

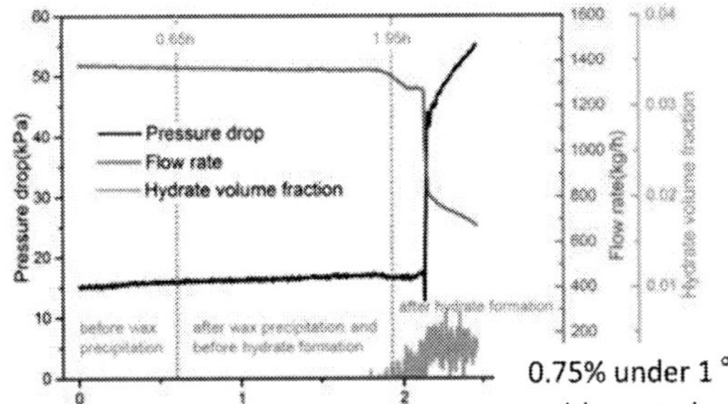


0.75% under 1 °C set temperature

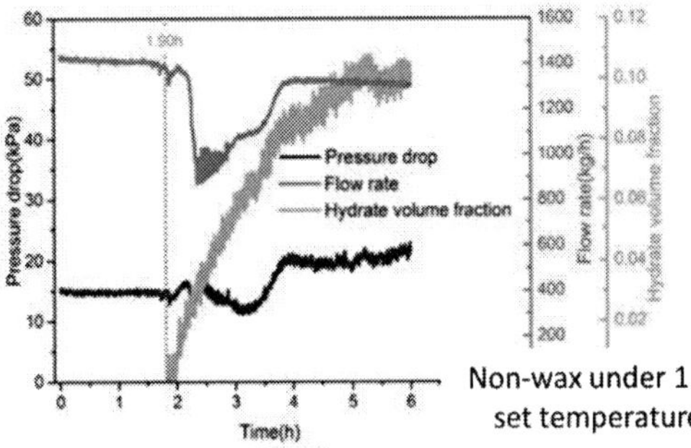


Non-wax under 1 °C set temperature

2017 AIChE ANNUAL MEETING

AIChE The Global Home of Chemical Engineers

## Results and discussion

- **Effects of wax on flow properties — Analyze**

$$h_l = \lambda \frac{v^2 L}{2gd} \qquad h_l = \beta \frac{Q^{2-m} \mu^m}{d^{5-m}} L$$

*for laminar flow, β=4.15 and m=1; for hydraulically smooth flow, β=0.0246 and m=0.25*

$$\Delta P = \Delta P(\text{vicosity}) + \Delta P(\text{flow diameter})$$

- Severe deposition
  → abrupt diameter decrease

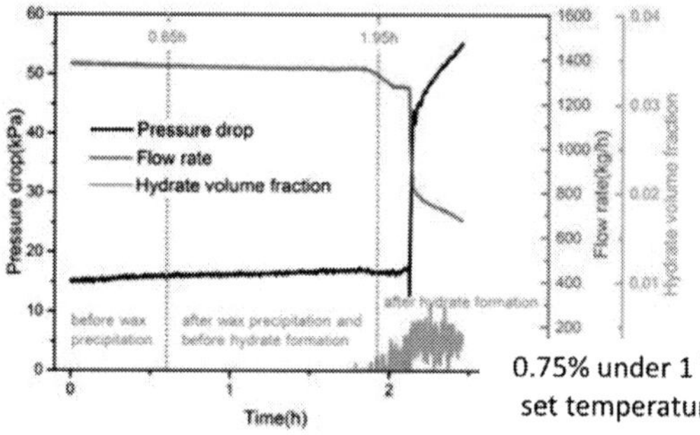


0.75% under 1 ℃ set temperature

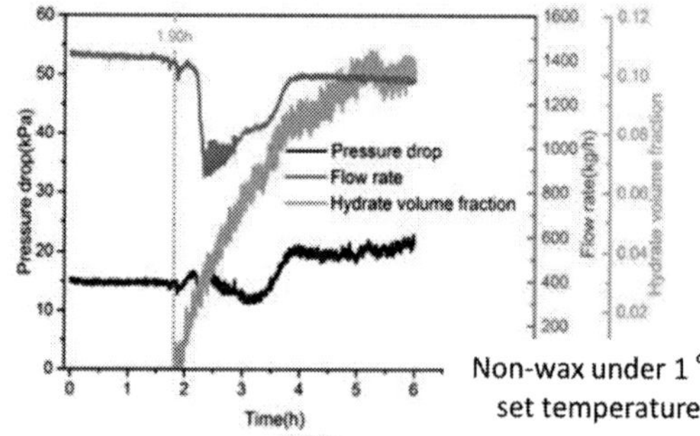


Non-wax under 1 ℃ set temperature

CHINA UNIVERSITY of PETROLEUM, BEIJING

8/18

2017 AIChE ANNUAL MEETING

AIChE The Global Home of Chemical Engineers

## Results and discussion

- **Effects of wax on flow properties — Analyze**
  - High pressure visual window

Before hydrate formation

5 mins after hydrate formation

9 mins after hydrate formation

15 mins after hydrate formation

⇒ Severe deposition process
→ rapid plugging

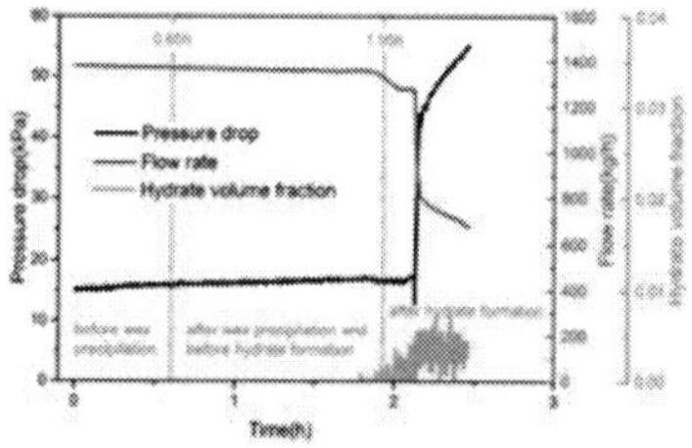


0.75% under 1 ℃ set temperature

Real time video

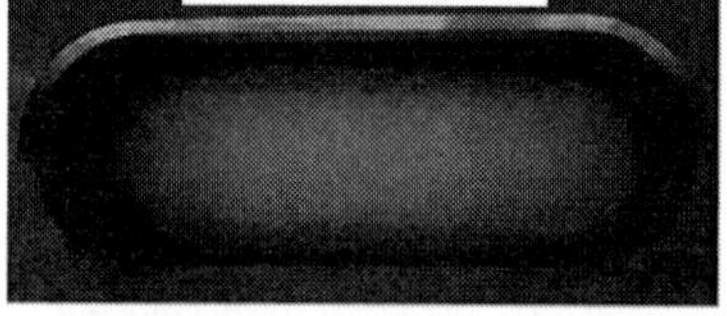

CHINA UNIVERSITY of PETROLEUM, BEIJING

9/18

## Results and discussion

- **Effects of wax on flow properties — Analyze**
  - PVM probe

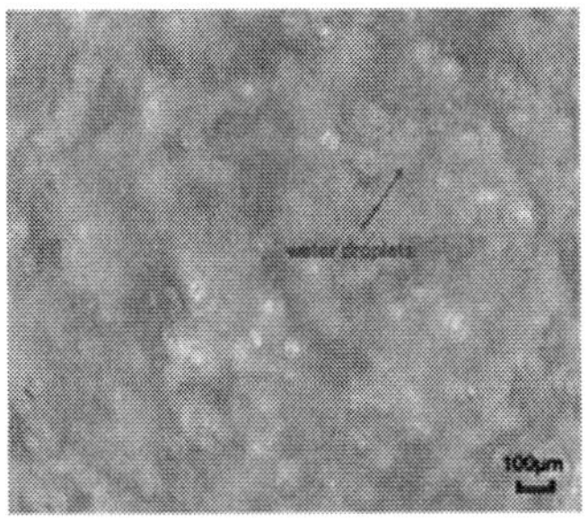


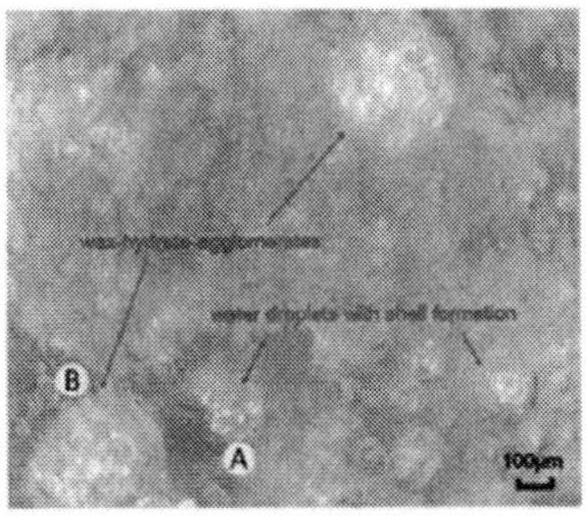


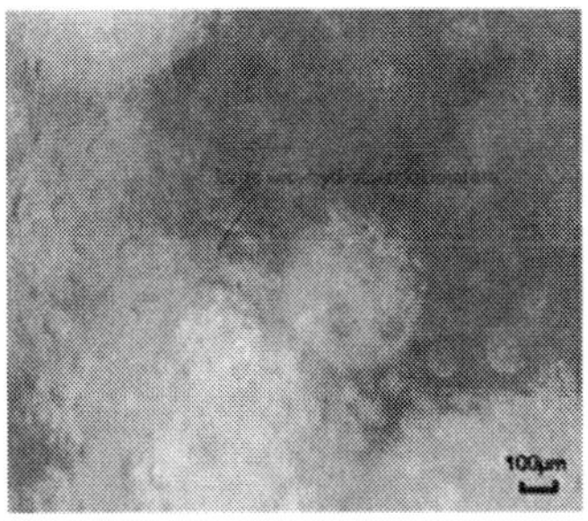


➡ Existence of wax-hydrate-agglomerates



2017 AIChE ANNUAL MEETING
AIChE
The Global Home of Chemical Engineers

## Results and discussion

- **Effects of wax on flow properties — Analyze**
  - FBRM probe

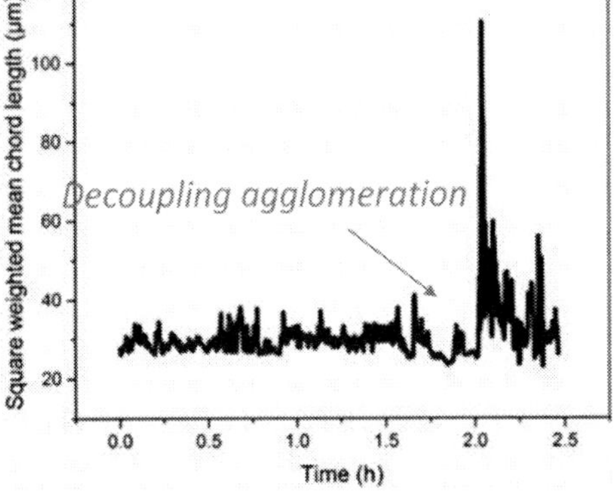


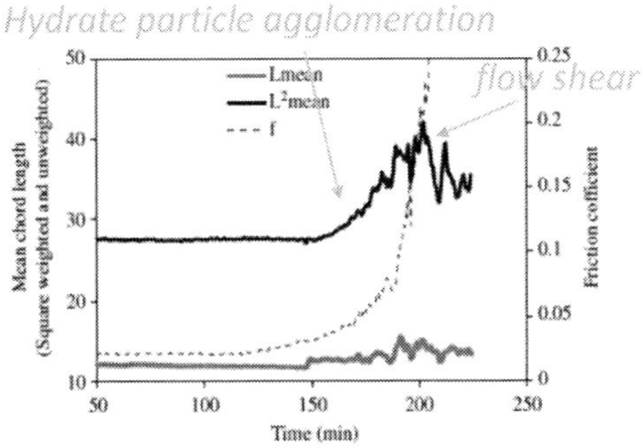


*Cited from H.Leba et al. Chemical Engineering Science 65(2010) 1185–1200*

➡ Deposition of wax-hydrate-agglomerates

## Results and discussion

- **Effects of different set temperatures**
  - Induction time

The lower the set temperature, the shorter the induction time.

Roughly linear relationships
→ driving forces

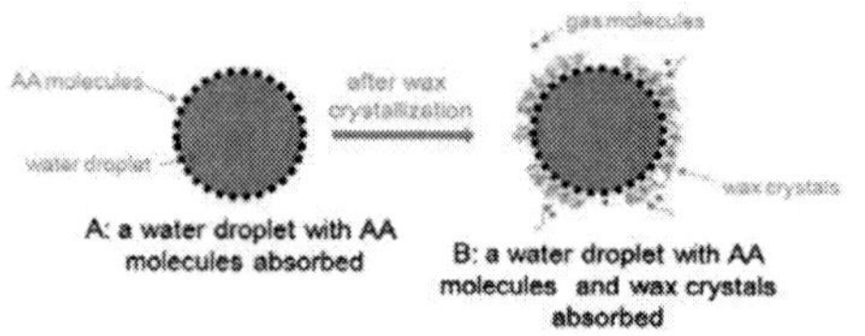


More wax crystals absorbed.

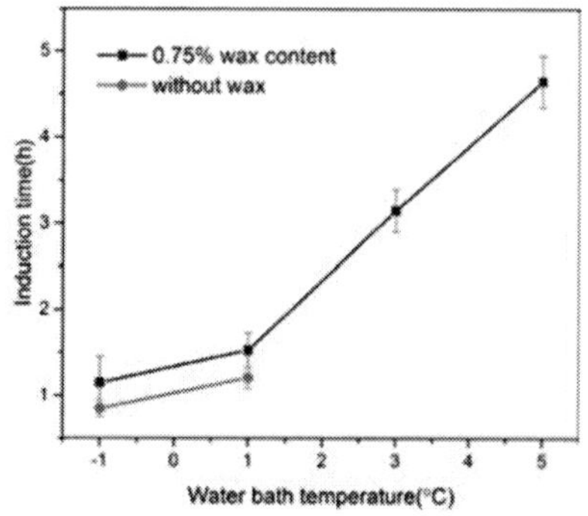


Induction times of
0.75% and non-wax



2017 AIChE ANNUAL MEETING
AIChE
The Global Home of Chemical Engineers

## Results and discussion

- **Effects of different set temperatures**
  - Flow properties

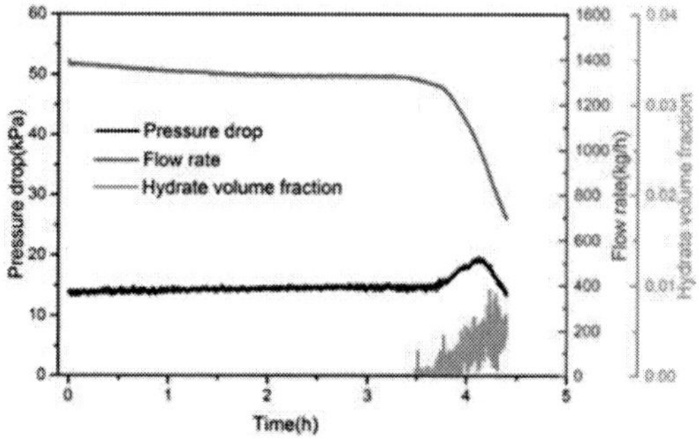


Set temperature= 3°C

Moderate sate

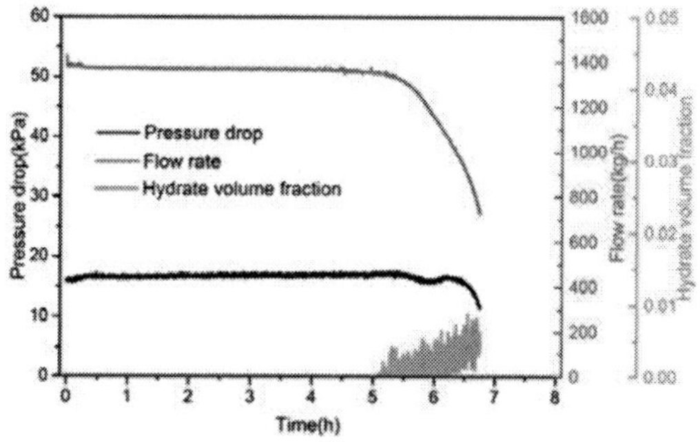


Set temperature= 5°C

Gradual increase of viscosity

2017 AIChE ANNUAL MEETING

AIChE
The Global Home of Chemical Engineers

## Results and discussion

- **Effects of different set temperatures**
  - Flow properties

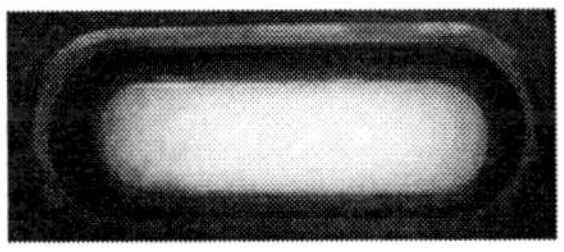

15 mins after hydrate formation

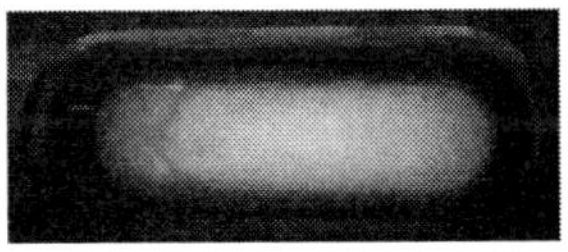

45 mins after hydrate formation

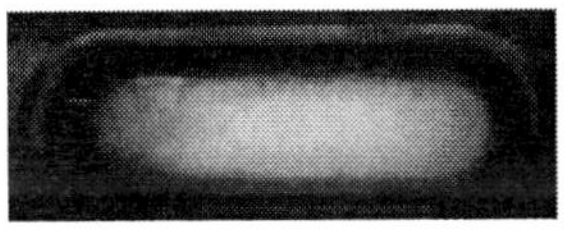

75 mins after hydrate formation

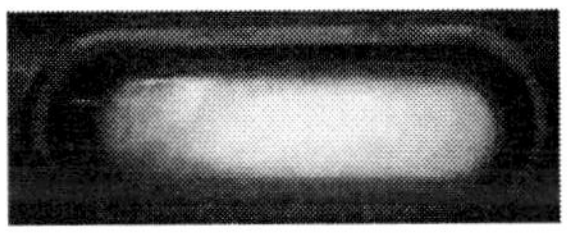

90 mins after hydrate formation

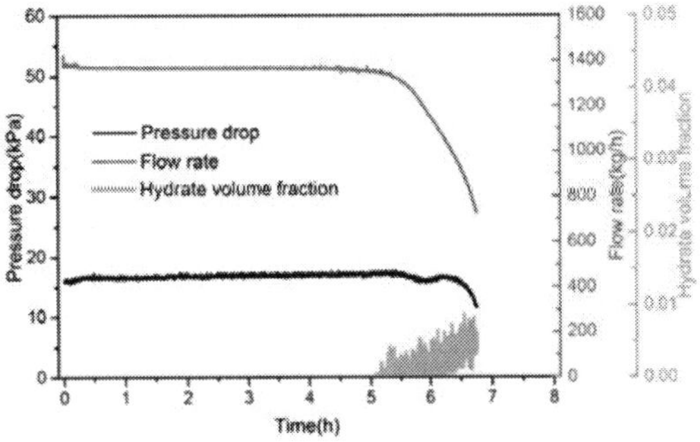


Set temperature= 5℃

Gradual increase of viscosity

No severe deposition happened
gradual deposition

CHINA UNIVERSITY of PETROLEUM, BEIJING

14/18

2017 AIChE ANNUAL MEETING

AIChE
The Global Home of Chemical Engineers

## Results and discussion

- **Effects of different set temperatures**
  - Flow properties

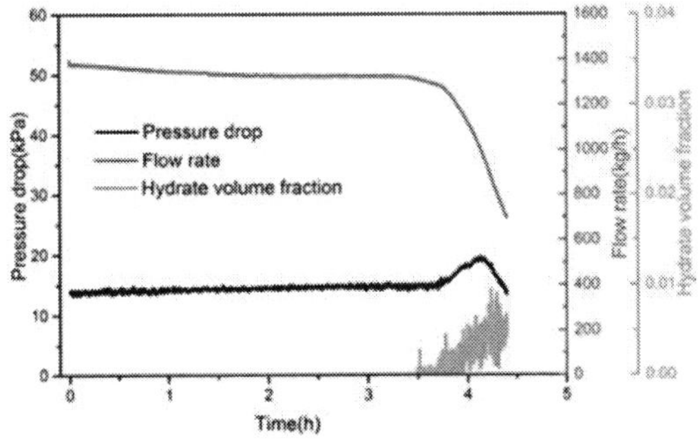


Set temperature= 3℃

Moderate sate

9 mins after hydrate formation

15 mins after hydrate formation

30 mins after hydrate formation

45 mins after hydrate formation

Moderate deposition sate

CHINA UNIVERSITY of PETROLEUM, BEIJING

15/18

## Conclusion

- An 80vol% diesel oil + natural gas + 20vol% deionized water + 1wt% AA + 0.75wt% wax experimental system was used to study the effect of wax on **hydrate formation induction time** in a water-in-oil system, and **flow properties** before and after hydrate formation.
- Hydrate formation induction time was prolonged by wax crystals due to **deposition wax** and **adsorbed wax crystals**.
- But once hydrate formed, a severe deposition process might happen due to **wax-hydrate-agglomeration**, resulting in rapid plugging.
- It's important to acquire the composition of products form wellheads and field engineers should be aware of this situation.



2017 AIChE ANNUAL MEETING
AIChE
The Global Home of Chemical Engineers

## Acknowledgement

- **National Natural Science Foundation of China** (No. 51534007)
- **National Key Research and Development Plan** (No. 2016YFC0303704)
- **National Science & Technology Major Project** (No. 2016ZX05028004-001 and 2016ZX05066005-001 ).
- **Science Foundation of China University of petroleum-Beijing** (No. C201602)

**China University of Petroleum-Beijing**

**National Engineering Laboratory for Pipeline Safety**

**Please Contact: ydgj@cup.edu.cn**

2017 AIChE ANNUAL MEETING
AIChE
The Global Home of Chemical Engineers
Thank you!
These results are going to be published soon!
CHINA UNIVERSITY of PETROLEUM, BEIJING
18/18

# 503670 Digital Alchemy for Assembly Engineeringc

**Abstract:**

Back in the day, Stuart Churchill was a pioneer in the use of computers for modeling chemical and thermal processes. Since that time, the role of computing in chemical engineering has become indispensable. In this talk, we present new computational methods we call "digital alchemy" that allow for predictive modeling of thermodynamic assembly, and permit the design of building blocks optimized for assembly into target structures with desired properties. In this approach, standard thermodynamic ensembles are generalized to include alchemical variables and their thermodynamic conjugates, allowing for rapid, real-time computational investigation of large, complex design spaces.

# 503671 Improved Algebraic, Numerical, and Graphical Representations in Fluid Mechanics

**Abstract:**

The concept of equivalents is applied to fully developed flows through straight channels, and two applications are introduced: the equivalents that produce equal rates of flow through channels of different cross-section, and the diminishment that occurs between the regimes of laminar and turbulent flow. The validity of the eddy diffusivity and the invalidity of the mixing-length as measures of the shear stress due to turbulence have perhaps never before been demonstrated so clearly. The possible choices of primary variables and the consequences are also examined. The objective of these explorations is an improved understanding of the basic fluid mechanics by teachers, students, and industrial practitioners.

# 503672 The Scaling of Turbulence Near the Wall and the Churchill Turbulent Flux Correlation: Insights with Lagrangian Simulations

**Abstract:**

The evaluation of mean velocity and mean temperature in wall-bounded turbulent flows has been based on the "classical" theory of turbulence and the law of the wall. The viscous wall parameters, namely the friction velocity, the friction temperature, and the kinematic viscosity of the fluid have been used to scale not only the mean values of velocity, but also other mean turbulent quantities close to the wall, including the Reynolds stresses and the turbulent velocity and temperature fluctuations. Stuart Churchill, as he questioned the validity of the Reynolds analogy between momentum and heat or mass transfer in his works [1], he realized that extending the scaling of momentum to the scaling of scalar quantities in turbulence was not a sound approach. He proposed a scaling of turbulence that depends on the turbulent momentum flux and the turbulent heat flux. In his AIChE Institute Lecture (in Miami, 1998), Churchill presented a summary of his work on this topic, and closed form solutions for the mean velocity and temperature [2]. The turbulent Prandtl number, the most significant parameter for the modeling of turbulent transport not only in theory but also in modern CFD commercial software packages, was also interpreted physically based on these ideas [3]. We have used direct numerical simulation results and results from Lagrangian simulations [4] for the transport of passive scalars in turbulent channel flows to explore Churchill's ideas on turbulent flow scaling and to develop an updated theory for the scaling of heat transport in turbulent flows [5-7]. His passion to remove empiricisms and approximations from the development of predictive correlations in turbulent transport has been an inspiration. In this talk, we will revisit the Churchill scaling in view of our simulation results, and will discuss the counterintuitive finding that wall turbulence can be used to separate rather than mix substances under special circumstances [8].

1. Churchill, S. W., 1996, "A Critique of Predictive and Correlative Models for Turbulent Flow and Convection," Ind. Eng. Chem. Res., 35, pp. 3122–3140.
2. Churchill, S.W., 2000, "Progress in the Thermal Sciences: AICHE Institute Lecture," AIChE Journal, 46(9), pp. 1704-1722.
3. Churchill, S. W., 2002, "A reinterpretation of the Turbulent Prandtl Number," Ind. Eng. Chem. Res., 41, pp. 6393–6401.
4. Papavassiliou, D.V, 2002, "Turbulent transport from continuous sources at the wall of a channel," Int. J. Heat Mass Transfer, 45(17), pp. 3571-3583.
5. Le, P.M., and Papavassiliou, D.V., 2008, "On the scaling of heat transfer using thermal flux gradients for fully developed turbulent channel and Couette flows," Int. Commun. Heat Mass Transfer, 35(4), pp. 404-412.
6. Srinivasan, C., and Papavassiliou, D. V., 2011, "Prediction of Turbulent Prandtl Number in Wall Flows With Lagrangian Simulations," Ind. Eng. Chem. Res., 50(15), pp. 8881–8891.
7. Srinivasan, C., and Papavassiliou, D.V., 2013, "Heat transfer scaling for wall bounded turbulent flows," Applied Mechanics Review 65(3), Art. 031002.
8. Nguyen, Q., Srinivasan, C., and Papavassiliou, D.V., 2015, "Flow induced separation in wall turbulence", Phys Rev E, 91, Art. 033019.

# 503674 Flow Boiling Using a Piranha Pin Fin Heat Sink

**Abstract:**

**Flow Boiling using a Piranha Pin Fin Heat Sink**

Corey Woodcock, Xiangfei Yu, Yoav Peles[1] and Joel Plawsky

Rensselaer Polytechnic Institute; [1]University of Central Florida

The *Piranha Pin Fin* (*PPF*) microdevice is a heat exchanger engineered to dissipate extreme thermal loads using either dielectric fluids such as HFE7000 or water. Fabrication of the device is CMOS compatible and the first generations of the device have been fashioned from silicon substrates. The PPF operates on the principle that heat transfer would be more efficient if heated fluids could be removed nearly as fast as they are created. This process would keep the driving force as high as possible. Thus the heat exchanger is built as two layers; a primary substrate that dissipates the bulk of the heat and shunts the heated fluids to a booster section that continues to heat the fluid and extends the total surface area of the device. Three generations of devices were fabricated. The first two generations were single-layer systems designed to help understand the operation of the primary layer and optimize the shape and location of the pin fins. The third generation was the first, two-layer system to be fabricated and tested.

Figure 1 Representative PPF array

Correlations for the heat transfer coefficient, $h_{tp}$, based on the Boiling Number and the Weber Number in the PPF design was developed and validated with Gen II. The discrepancy of heat transfer coefficient between predicted and measured values was within 20%. The ratio of two-phase to single-phase pressure drop was correlated with the Boiling Number (Bo) and the Jakob Number (Ja). The discrepancy of the two-phase multiplier between prediction and experimental measurements were within 12%.

The third generation *PPF* heat sinks were the first series to provide both a primary substrate populated with *PPF*s, as well as a booster substrate populated with solid micro pin fins. The substrates are micro-aligned and fusion-bonded to provide a full three dimensional (3D) architecture with bulk silicon properties. Experimental results from the Gen III *PPF* microdevices provide a proof of concept of the true *PPF* heat sink concept with fluid venting into multiple substrates. At low flowrates ($G$ = 670 [kg/s/m$^2$]) the Gen III *PPF* heat sink has been shown to dissipate heat loads approaching 1.1 [kW/cm$^2$] at an exit quality of 95%, while at higher flowrates ($G$ = 3,300 [kg/s/m$^2$]) the thermal limit of the heat sink remains unknown due to exceeding the available power supply at >1.4 [kW/cm$^2$].

# 503673 Transport Problems in the Spirit of Stuart Churchill for Teaching and Research at the University of Michigan

**Abstract:**

Stuart Churchillâ€™s solutions to important heat transport problems have greatly influence both pedagogy and research. The correlation of Churchill and Chu for the natural convection heat transfer to or from a vertical plate, over an astonishing range of Rayleigh numbers is based partly on a similarity solution of a type common in heat, mass, and momentum transfer for problems in chemical engineering. One such problem is asphaltene or wax transport and deposition onto the wall of an oil pipeline, which can be solved using similar methods, as will be discussed. In addition, such problems, and the wealth of their applications make for excellent motivations and examples in the teaching of graduate or undergraduate transport, as will be discussed in the talk.

# Multiphase equilibrium calculation framework for compositional simulation of $CO_2$ injection in low temperature reservoirs

Huanquan Pan
Michael Connolly
Hamdi Tchelepi

Energy Resources Engineering Department
Stanford University

## Outline

- Introduction
- Numerical algorithms
- Initial estimates for phase stability test calculations
- Results for calculation of phase diagrams
- Results for numerical algorithms
- Conclusions

## Introduction

- $CO_2$ EOR (enhanced oil recovery) has been operated for many decades in Texas and New Mexico of USA.
- Interest rises in $CO_2$ EOR for carbon storage and sequestration.
- Three-hydrocarbon phases (V: vapor, $L_1$: oil-rich liquid, $L_2$: $CO_2$-rich liquid) form at high $CO_2$ concentrations when temperature T < $120^0$F.
- Complex displacement mechanism requires full three-hydrocarbon-phase compositional simulation at field scale. No such simulator exists.
- Three-hydrocarbon-phase simulator requires robust and efficient multiphase (up to 3) equilibrium (stability testing and flash) calculations.



## Research goal

- Multiphase equilibrium calculations for three-phase hydrocarbon-$CO_2$ systems have been studied over three decades
- Existing techniques are unable to calculate the entire three-phase region
- Goal: to develop a framework for robust and efficient calculations of multiphase equilibrium in the simulation of $CO_2$ EOR in low temperature reservoirs

# Step-wise three-phase equilibrium calculations

Given overall composition, T and P, determine if system is one, two or three phases and corresponding phase amounts and compositions

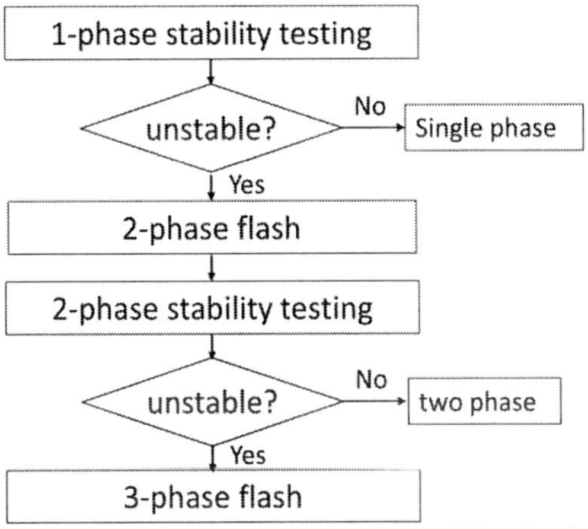




# Challenges in multiphase equilibrium calculations

- Problems need to be solved
  - 1-phase stability testing
  - 2-phase stability testing
  - 2-phase flash
  - 3-phase flash
- To solve each problem
  - Numerical iteration algorithm
  - Initial values to start iteration
    - Initial values for flash are from stability testing
- Challenges for multiphase equilibrium calculations
  1. Efficient and robust algorithm to handle the difficult conditions in stability test and flash calculations
  2. Efficient initial estimates for stability test calculations

## Review of algorithms for phase equilibrium calculations

- SSI (successive substitution iteration) + Newton
  - Very efficient
  - Encounter highly difficult conditions in both stability test and flash
  - Not robust enough for compositional simulation
- Trust-region optimization
  - Globally convergent, guarantees convergence
  - Computational cost is higher than Newton
  - Used only when both SSI and Newton methods encounter difficulty



## Proposed algorithm for phase equilibrium calculations

We implemented SSI-Newton-trust region optimization algorithm to calculate phase-stability testing and multiphase flash.

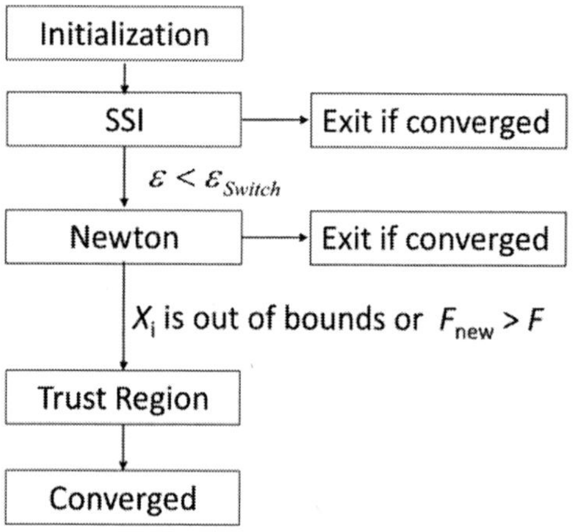

## Formulation of stability test calculation

Given composition $z_i$, minimize TPD (tangent plane distance)

$$TPD = 1 + \sum_{i=1}^{Nc} Y_i \left( \ln Y_i + \ln \hat{\phi}_i^V (Y_i) - \ln z_i - \ln \hat{\phi}_i^L (z_i) - 1 \right)$$

with respect to $Y_i$ (mole number of new phase).

$\sum_{i=1}^{N} Y_i^* < 1$ Stable; $\sum_{i=1}^{N} Y_i^* > 1$ Unstable



## Formulation of multiphase flash calculation

Given composition $z_i$, minimize Gibbs free energy $\overline{G}$ with respect to $n_i^p$ (mole number of component i in phase p; i=1, ..,Nc; p=1, ..., Np-1)

$$\overline{G} = G / RT = \sum_{i=1}^{Nc} \sum_{p=1}^{Np} \left( n_i^p \ln \hat{f}_i^p (n_i^p) \right)$$

Subject to material balance with feed composition $z_i$

$$\sum_{p=1}^{Np} n_i^p = z_i$$

Total independent variables: $Nc \times (Np - 1)$

## Solve Rachford-Rice (RR) equations for three phases

RR equations are solved to get phase moles (*V* and $L_2$) for each SSI by giving phase equilibrium constants $K_i^V$ and $K_i^{L_2}$

$$\sum_{i=1}^{Nc} \frac{z_i(K_i^V - 1)}{1 + V(K_i^V - 1) + L_2(K_i^{L_2} - 1)} = 0 \qquad \sum_{i=1}^{Nc} \frac{z_i(K_i^{L_2} - 1)}{1 + V(K_i^V - 1) + L_2(K_i^{L_2} - 1)} = 0$$

RR equations are highly non-linear and very good initial *V* and $L_2$ values are required to solve them with Newton iterations.
We use Trust-Region to minimize the following function to get *V* and $L_2$

$$F = \sum_{i=1}^{Nc} -z_i \ln\left[1 + V(K_i^V - 1) + L_2(K_i^{L_2} - 1)\right]$$

- Globally convergent with any initial V and $L_2$ values, 1/3 is default estimate for V and $L_2$
- Initial and converged V and $L_2$ can be negative values



## Review of initial estimates for stability testing

Initial estimate of phase-equilibrium constant $K_i$-values is required to start stability test calculation. Since local optimization method is used in stability testing calculation, multiple initial estimates are required to ensure phase instability is detected

Proposed initial estimates in literature:

$$K_i^{Wilson},\ 1/K_i^{Wilson},\ (K_i^{Wilson})^{1/3},\ 1/(K_i^{Wilson})^{1/3},\ K_i^{pure-j}\ (j = 1,2,\ldots,Nc)$$

$$K_i^{Wilson} = \frac{P_{Ci}}{P}\exp\left[5.37(1+\omega_i)(1 - T_{Ci}/T)\right]$$

$$K_j^{pure-j} = 0.9/z_i\,,\ K_i^{pure-j} = 0.1/(Nc-1)/z_i\ \ (j \neq i, i = 1,\ldots,Nc)$$

Total: Nc+4 estimates

Li & Firoozabadi (2013)

## Review of initial estimates for stability testing

- Existing approach
  - Use all Nc+4 estimates for single-phase stability testing
  - Assume that only one phase is required to be tested for a two-phase system in two-phase stability testing
  - Recommend testing only heavy liquid phase
  - Use all Nc+4 estimates for two-phase stability testing

- Outstanding issues
  - Too many estimates [total 2(Nc+4)], most of them generate the same converged TPD results
  - Fail to detect a lot of points in three-phase region if only heavy phase is tested
  - Existing initial estimates (Nc+4) cannot detect the instability in entire three-phase region



## Proposed sets of initial estimates

- 1-phase stability testing: 5 estimates

$$K_i^{Wilson},\ 1/K_i^{Wilson},\ K_i^{CO_2},\left(K_i^{Wilson}\right)^{1/3},\ 1/\left(K_i^{Wilson}\right)^{1/3}$$

- 2-phase stability testing:
  - Proposed new estimate:

$$K_{CO_2}^{CO_2-95}=0.95/z_{CO_2},\ K_l^{CO_2-95}=0.05/(Nc-1)/z_l\ \ (l\neq CO_2, l=1,\ldots,Nc)$$

  - Set of estimates:
    - Test heavy phase $K_i^{Wilson},\ K_i^{CO_2},\left(K_i^{Wilson}\right)^{1/3},\ K_i^{CO_2-95}$
    - Test light phase $K_i^{CO_2},\ 1/\left(K_i^{Wilson}\right)^{1/3},\ K_i^{CO_2-95}$
  - Total 7 calculations for 2-phase stability testing.

## Number of calculations in stability testing

Compare with existing method in literature:

- 1-phase calculations: 5 vs. Nc+4
- 2-phase calculations: 7 vs. Nc+4
  (Nc is typically between 6 to 10)



## Test fluids

Nine fluids, most of them are from West Texas or New Mexico of USA.

| Fluid | BSB | JEMA | NWE | OilG | MRO | MSO | OilB | ResA | ResB |
|---|---|---|---|---|---|---|---|---|---|
| T (K) | 313.71 | 316.48 | 301.48 | 307.59 | 305.35 | 305.35 | 307.6 | 311.48 | 316.48 |

Data of thermodynamic properties:

- Khan et al., SPE24130, 1992. BSB (Bob Slughter Block), JEMA, NWE (North Ward Estes), OilG.
- Li & Firoozabadi, SPE J. 17, 1096-1107, 2012. MRO (Maljamar Reservoir Oil), MSO (Maljamar Separator Oil), OilB
- Pan et al., SPE REE, 18, 250-263, 2015. ResA and ResB

## Summary of tests

- Constructed P (pressure, bar) – $x_{CO2}$ (injected $CO_2$ mole fraction in mixture) phase diagram in 1-phase, 2-phase and 3-phase regions
- Used high-resolution sampling
  - 3-phase region: $\Delta P$=0.01 or 0.02, $\Delta x$ = 0.001; 250,000 to 650,000 test points for each fluid
  - 1-phase or 2-phase region: $\Delta P$ =1, $\Delta x$ =0.001; around 100,000 test points for each fluid
- Tested around 5 million points with many repeated runs; not a single failure encountered (e.g. divergence, stopping execution, et al.)



## Outline

- Introduction
- Numerical algorithms
- Initial estimates phase stability testing
- Results for calculation of phase diagrams
- Results for numerical algorithms
- Conclusions

# Results from existing approach (Nc+4 initial estimates)

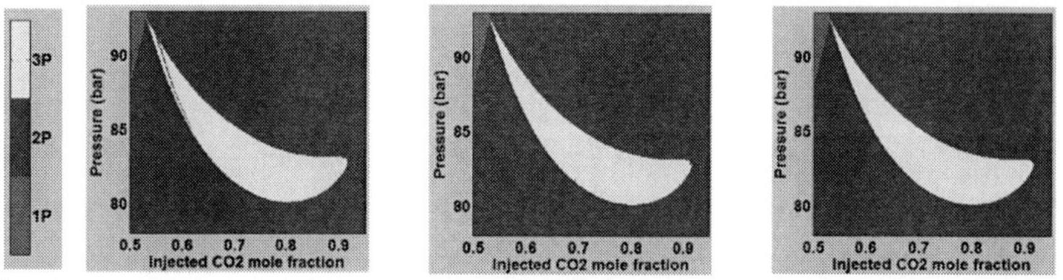


BSB fluid: a. Test heavy phase b. Test light phase c. Test both phases

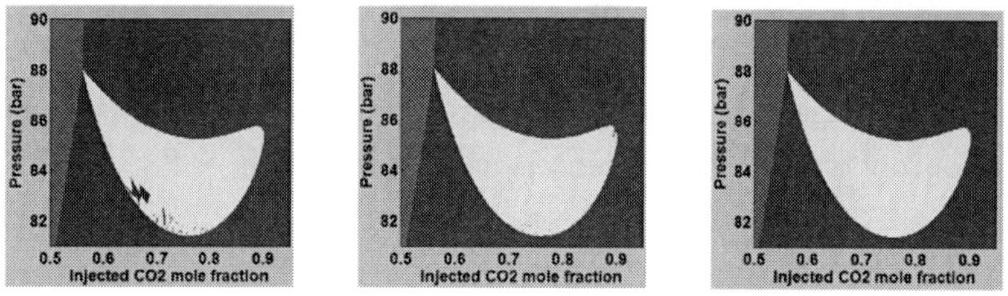


JEMA fluid: a. Test heavy phase b. Test light phase c. Test both phases



# Results from existing Nc+4 initial estimates

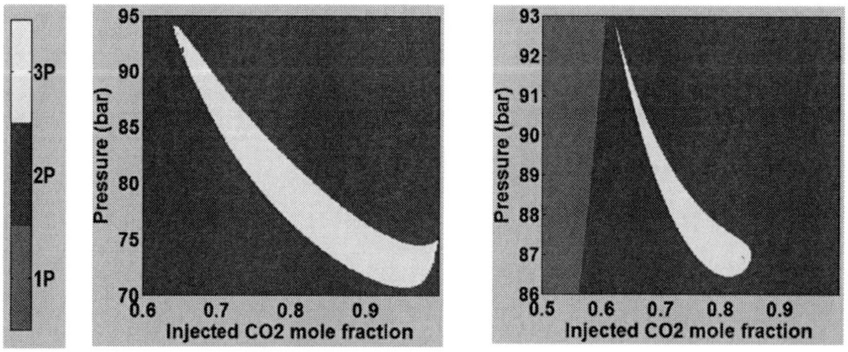


MRO ResB

**P-X diagram without using initial $K_i^{CO_2-95}$ estimate**

## Calculated phase diagrams in our approach

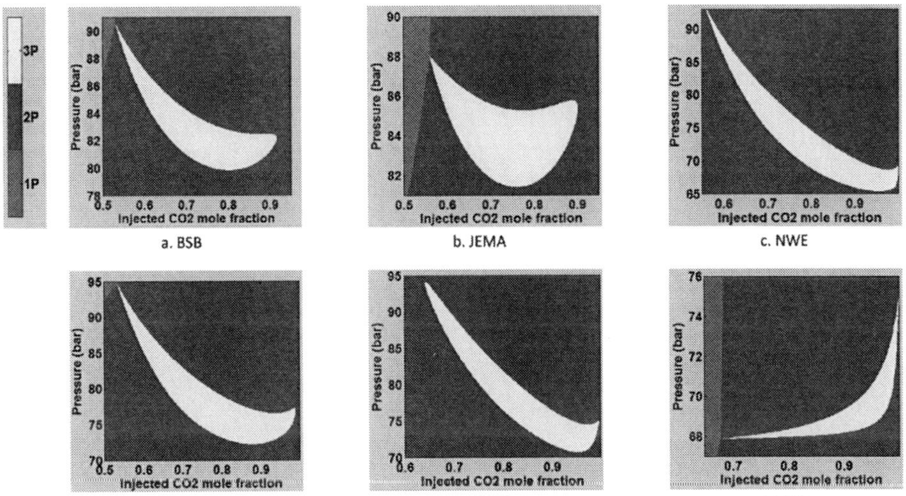


a. BSB b. JEMA c. NWE
d. OilG e. MRO f. MSO



## Calculated phase diagrams in our approach

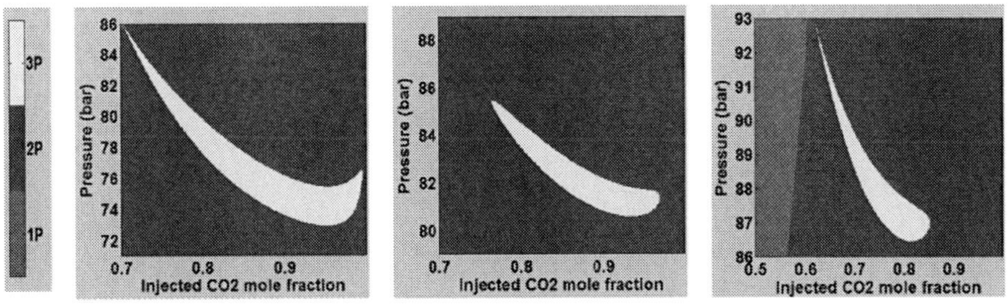


g. OilB h. ResA i. ResB

## Outline

- Introduction
- Numerical algorithms
- Initial estimates phase stability testing
- Results for calculation of phase diagrams
- Results for numerical algorithms
- Conclusions



## Iterations for stability test calculations

2-phase stability testing at $X_{CO2}$=0.603, P=82.4 bar for OilG

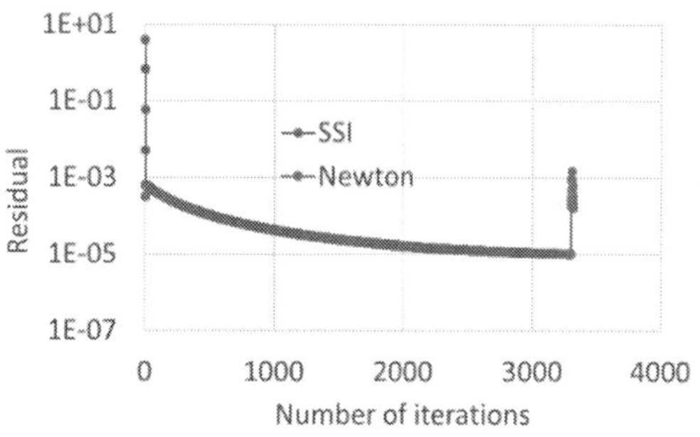


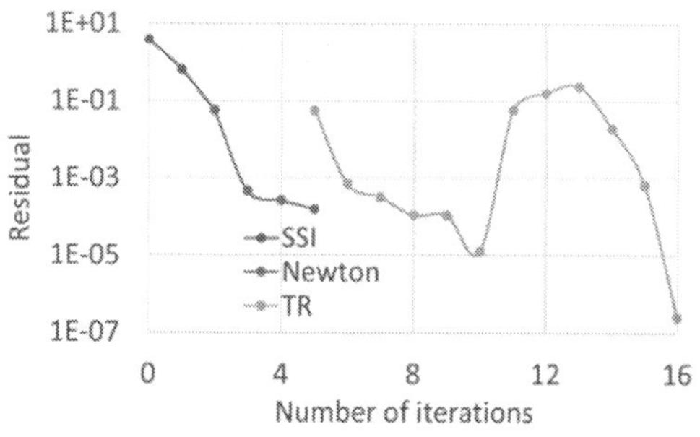


SSI-Newton

- Switch residual: $10^{-5}$

SSI-Newton-Trust Region

# Iterations for 2-phase flash calculation

$X_{CO2}$=0.747, P=77.06 bar for OilG

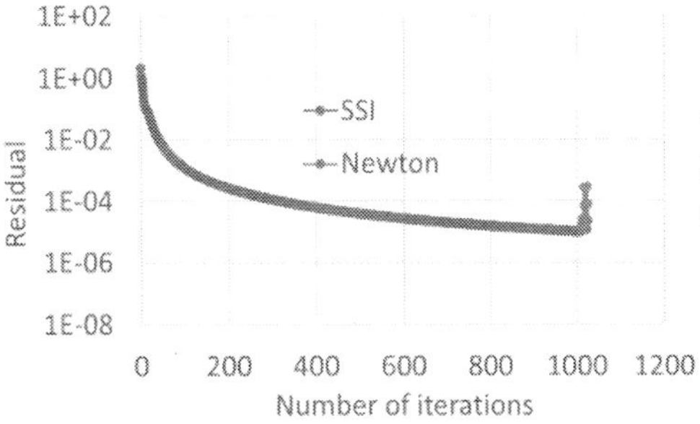


SSI-Newton

- Switch residual: $10^{-5}$

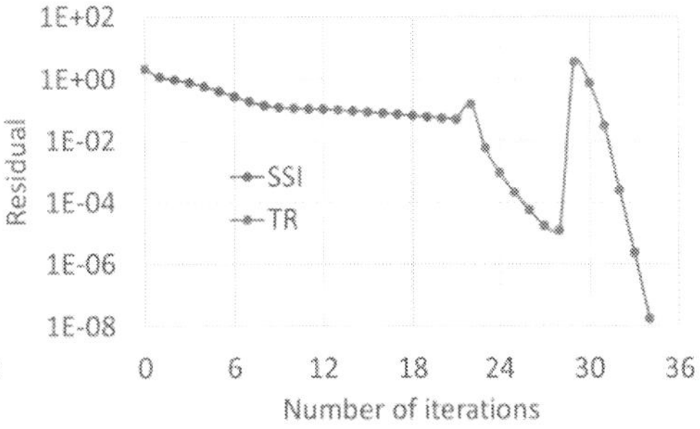


SSI-Newton-Trust Region (T-R)



# Iterations for 3-phase flash calculations near bi-critical point

$X_{CO2}$=0. 893, P=85.72 bar for JEMA

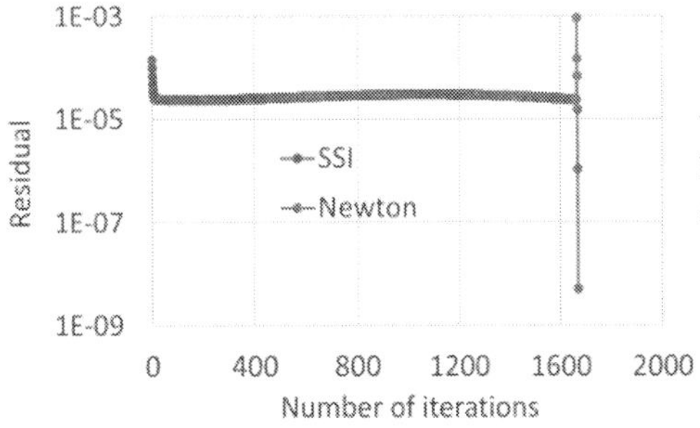


SSI-Newton

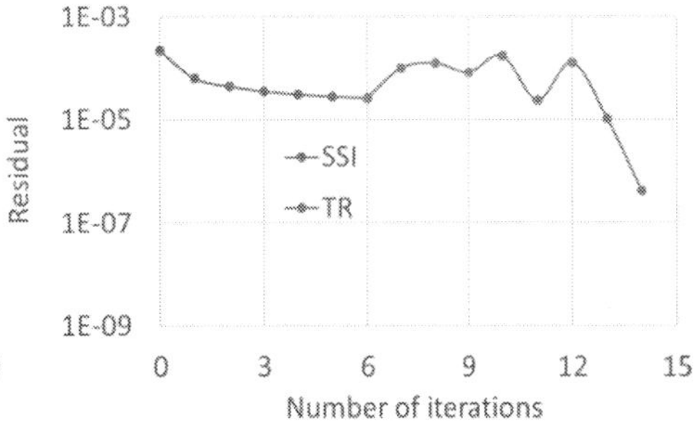


SSI-Newton-Trust Region (T-R)

Newton converges only when starting residual <2.34x$10^{-5}$ after more than 1500 SSI

# Iterations for solving Rachford-Rice equations

*$K_i$-values are* from 2-phase stability testing and flash at $X_{CO2}$=0. 893, P=85.72 bar for JEMA

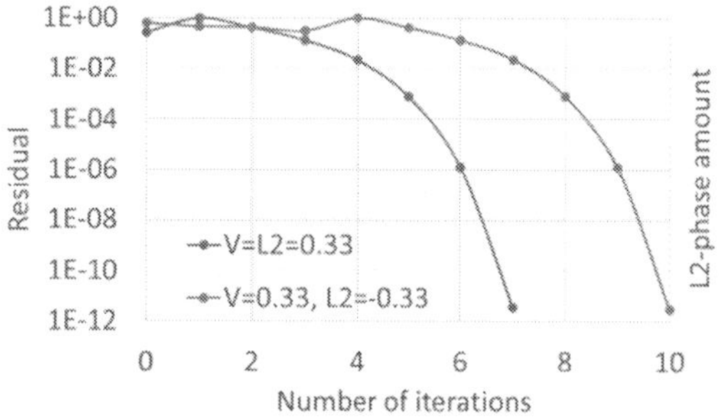


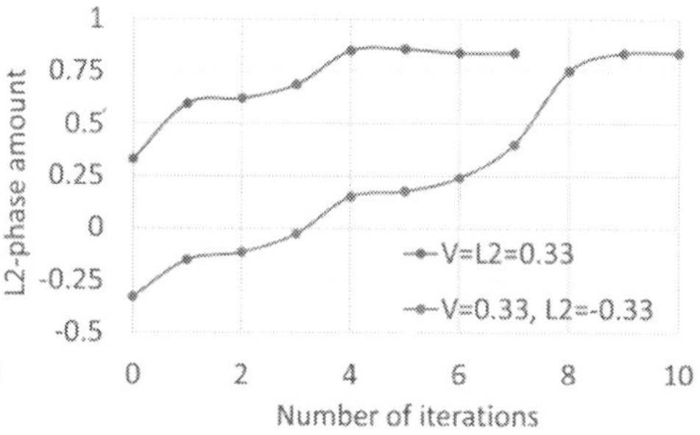


Large number of iterations only incurs in the first SSI. After first SSI, initial V and $L_2$ are from previous SSI, only one or two iterations.



# Conclusions

- Implemented sequential iterations of SSI-Newton-Trust Region algorithm for robust and efficient calculations in phase stability testing and multiphase flash
- Developed efficient trust-region based solver for Rachford-Rice equations
- Proposed a new initial estimate for phase-stability testing, optimized the sets of initial estimates for 1-phase and 2-phase stability testing to reduce the computational cost of stability testing

## Conclusions

- Demonstrated the necessity of testing both phases, rather than testing one phase in 2-phase stability testing
- Demonstrated the efficiency and robustness of the computational framework by extensively and rigorously testing millions of points with nine reservoir fluids
- Framework developed in this research meets the requirements for multiphase equilibrium calculations in the compositional simulation of $CO_2$ injection in low temperature reservoirs

# 491867 Phase Behavior in Nanoporous Media and Its Impact on Hydrocarbon Recovery

**Abstract:**

It has been well recognized that production from liquid-rich shale (LRS) reservoirs behaves very differently from conventional reservoirs, and one of the potential causes has been widely attributed to effects of poreâ€"confinement on phase behavior and mass transfer kinetics. The underlying physics behind multiphase fluid flow in unconventional tight reservoir has not yet been fully understood, and the effects of pore-confinement have been mostly assessed by numerical modeling/simulation in the literature.

This paper reports our laboratory measurements of hydrocarbon phase transition and mass transfer kinetics in nano-pore media for both single-component and multi-component hydrocarbon systems. The experimental results confirmed the significant impact of pore-confinement on gas/oil phase transition and flow in nano-pores, and the multiphase flow dynamic is examined to elucidate the fundamental physics involved. This offers a more realistic description of pertinent mechanisms for hydrocarbon recovery from tight LRS formations and provides data for validating numerical modeling methodologies, which will enable us to predict and optimize production performance of LRS reservoirs.

# 500275 Ultra-Small-Angle Scattering Investigation of the Structure and Self-Assembly Mechanism of Asphaltenes in Solvent Mixtures

**Abstract:**

It is known that molecules of petroleum asphaltenes can self-assemble in crude oil and solvents into fractal clusters with Rg<40 nm and fractal dimensions ranging between 1.6 and 2.1 [1]. With the addition of an antisolvent (e.g., heptane), the stable clusters of asphaltenes can be destabilized and further aggregate into insoluble clusters with sizes in the micrometer length scale and a more compact internal structure (i.e., higher fractal dimension) [1]. However, it is still unclear what structural changes occur at the nano and micrometer length scales for both soluble and insoluble asphaltenes during the destabilization and precipitation process. In this talk, ultra-small-angle X-ray scattering (USAXS) results generated on beamline 9-ID-C at the Advanced Photon Source at Argonne National Lab will be presented to explore the mechanism of asphaltene destabilization, self-assembly and precipitation. Because of the ability of USAXS to be able to probe a wider range of scattering vectors (q), both soluble and precipitating asphaltenes with larger length scales can be detected simultaneously. The results show the capability of USAXS to track time-resolved changes in structural properties of asphaltenes as they transit from the nano to the micrometer length scale. The precipitation process was monitored as a function of time after adding an antisolvent by measuring the structural changes of both soluble and insoluble/precipitating clusters for long mixing times (>24 hours) and large length scales (>1000 nm). It was observed that the compactness and size of insoluble clusters increase as a function of time, while the structural properties of soluble asphaltenes remain the same. Moreover, it is revealed that the precipitating asphaltene clusters have surface fractal geometry or rough surfaces, but with a compact core. Trace quantities of inorganic solids (0.1nm<R<1 micron) were detected and likely contributed to the precipitation mechanism by acting as nucleation sites. Inorganic solids are present in real-world oil production and processing environments, and this result is of direct scientific and industrial importance. Overall, this study provided an improved mechanistic understanding of asphaltene assembly during precipitation and showed the powerful ability of USAXS for the investigations of asphaltene-related problems.

Â

[1] Hoepfner, M. P.; Vilas Boas Favero, C.; Haji-Akbari, N.; Fogler, H. S. The Fractal Aggregation of Asphaltenes. Langmuir 2013, 29, 8799â€“8808.

# 497096 An Atomistic Investigation of Hydrocarbon Behavior Confined in Kerogen Nanopores

**Abstract:**

The kerogen (a.k.a. a solvent insoluble organic matter in shales) holds a significant amount of hydrocarbons in free, adsorbed and absorbed forms. The limited understanding of hydrocarbon behavior confined in the nanoporous kerogen could be a limiting factor in the primary migration of hydrocarbons in kerogen and eventually on the overall hydrocarbon recovery in unconventional resources. A systematic investigation is performed to understand the influence of the chemical heterogeneity (of kerogen) and the variability in nanometer/subnanometer pore sizes (in kerogen) on the behavior of $CH_4$ confined in kerogen pores. In particular, the $CH_4$ distributed in free, adsorbed and absorbed forms are quantified based on the pore sizes. Monte Carlo/Molecular Dynamics simulation tools were used for the analysis. The proposed approach shed light on the complex hydrocarbon distribution in confined kerogen pore system, which is crucial to estimate the hydrocarbon in place and fraction of the hydrocarbon that are recoverable. Further, the proposed approach can be readily extended for long chain hydrocarbons and have broad implications in enhanced hydrocarbon recovery from unconventional resources.

# 488314 Effect of Temperature Shift Around Critical Point on Liquid Production from Shales

**Abstract:**

Light tight oil (LTO) production in US has seen a tremendous growth. The properties of crude oil such as density, viscosity etc. which directly affect the flow ability inside the porous rock depend on compositions of fluid and the reservoir conditions mainly temperature and pressure. The pressure in the reservoir declines in course of production as more hydrocarbons are extracted. A reservoir can be assumed to be isothermal at a smaller reservoir scale but variation in temperature can be observed in larger scale across same play such as in Eagle Ford Play in Texas, USA. Temperature also changes along the fluid flow pathway from sub-surface to surface facility. In this research, we investigated the effect of temperature changes in the sub-surface reservoir on thermodynamic fluid properties of hydrocarbons and consequently on their production. The thermodynamic fluid properties can be derived from its pressure-temperature diagram (PT- diagram). In petroleum industry, it is very common to determine the type of a fluid (dry gas, wet gas, gas-condensate, volatile oil and black oil) from initial reservoir pressure and temperature using the PT-diagram as shown in Figure 1. For same compositions, temperature is the deciding factor to determine the type of fluid reside in the reservoir. For instance, if the reservoir temperature is higher than critical point temperature, fluid is gas-condensate, wet gas or dry gas. On the other hand, temperature lower than critical point yields volatile oil and black oil. Volatile oil and gas-condensate are the two fluids which are the closest to the critical point but on the opposite sides (Figure 1).

figure 1: P-T plot showing variation in fluid properties with changing temperature and pressure of the reservoir

This fact causes the temperature change near critical point more sensitive. In this study, 25$^{O}$F temperature variation near critical point is considered to make gas-condensate (critical point+25$^{O}$F) and volatile oil (critical point-25$^{O}$F) for the same composition. Eagle Ford, a prolific liquid play in the USA is good example of this type of temperature variation. Volatile oil is initially liquid with large amount of gas dissolved in it and when pressure drops below bubble point pressure gas evolves out of liquid phase and causes two phases. In case of gas-condensate, fluid is initially gas containing large amount of volatilized oil, and eventually, the liquid hydrocarbon drops out from gaseous phase when pressure declines below dew point pressure and causes two phase of gas and liquid condensate (note that in case of gas-condensate, low pressure causes liquid to form, it is counter intuitive, that's why it is sometimes called retrograde condensate). It desired to produce volatile oil in liquid form and separate dissolved gas in the separator on the surface. In case of condensate, gas containing volatilized oil should be produced and condensate will be dropped out at separator by cooling and pressure drop. In reality, pressure goes below saturation pressure during production. Therefore, in either cases (volatile oil and gas-condensate) two phase forms inside reservoir when pressure goes below saturation pressure (bubble point or dew point pressure) during production. Gas being lighter and more mobile than liquid oil, always dominates the two phases or multiphase flow, suppressing the flow of liquid i.e., volatile oil and condensate. Therefore, production rate of liquid drops drastically when pressure drops below saturation pressure and a large amount of gas is produced along with volatile oil or condensate on the surface. In the current work, the recovery factor and liquid to gas ratio are mainly studied for five different initial compositions (fluid 1, fluid2, fluid3, fluid4 and fluid5). Each fluid is distinguished by different PT-diagram, critical point, saturation pressures, initial liquid to gas ratio (LGR) etc. Fluid 1 is the leanest in nature i.e, less oil and more gas (low LGR) and fluid 5 is the richest (high LGR). The effect of temperature variation (in volatile oil and gas condensate reservoir) for different reservoir permeabilities and operating pressures are studied. It was observed using reservoir simulations that a higher amount of liquid hydrocarbon was produced when temperature is higher than critical points i.e., when fluid is gas-condensate. Well could be operated at low bottom hole pressure in order to increase the total flow rate but it is not always preferable. It is found in this study that the higher bottom hole pressure (3000 psia) favors higher liquid recovery in 100 nD reservoir because of the fact that at lower bottom hole pressure (1000 psia) , gas dominates the two phase flow. Lower bottom hole pressure (1000 psia) is advantageous for higher permeability reservoir (1000 nD). Fluid with higher liquid content (such as fluid 4 and fluid 5) are less affected by the operating pressure. Recommendations based on reservoir scale simulation are given in this work for increased production at higher rates in shale plays.

17AIChE
Annual Meeting, Minneapolis, MN

ADNOC

# Wettability Analysis of Calcite Aged in Brine, Oil, and Smart Water

Imran K. Shaik[1], Prem K. Bikkina [1], Jin Song [2], Sibani L. Biswal [2], George J. Hirasaki [2], and Clint P. Aichele [1]

[1]Oklahoma State University, Stillwater, OK, USA

[2]Rice University, Houston, TX, USA

Phase Behavior and Flow of Reservoir Fluids

October 30, 2017

OKLAHOMA STATE UNIVERSITY

RICE
Unconventional Wisdom

# Objective

"To investigate wettability alteration mechanism(s) in carbonate reservoirs in order to improve oil recovery"

# What is Wettability?

- Wettability is the relative affinity of a solid substrate to a fluid in the presence of another immiscible or sparingly soluble fluid
- Contact angle is the measure of wettability

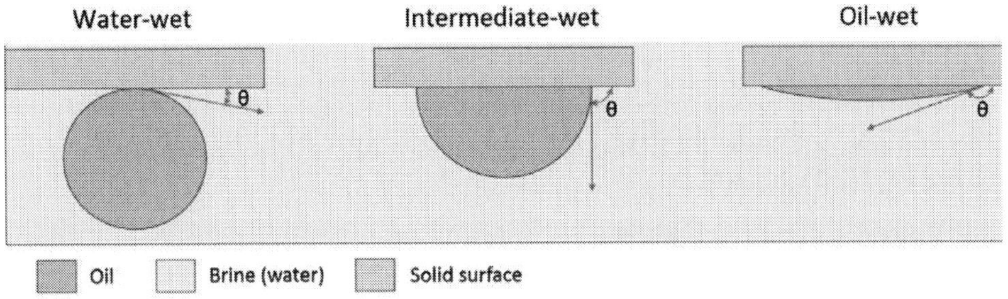


# Introduction

- Carbonate reservoirs
  - 60% of the world's oil
  - Very low oil recovery percentage during waterflooding
  - Naturally water-wet
  - Adsorption of carboxylic material make them oil-wet
- Wettability alteration
  - Towards water-wet can produce more oil
  - Tuning ionic composition of injected water (Smart water)
    - Decreasing salinity
    - Adjusting potential determining ions (PDIs) composition

# Materials

- Optical grade and polished Iceland Spar crystal as calcite surface
- 87:13 (v/v) mixture of n-Dodecane and Toluene as model oil
- Naphthenic acid as carboxylic material and acid number modifier
- ACS grade NaCl, $Na_2SO_4.10H_2O$, $MgCl_2.6H_2O$, and $MgSO_4.7H_2O$ salts
- DI Water with resistivity 18.2 MΩ cm

5

# Goniometer with Tilting Base

20 ° Tilt View

- Pressures up to 10,000 psig
- Temperatures up to 350 °F (177 °C)
- Kalrez® (FFKM) O-rings
- Tilting base
- Rame-hart's Dropimage Advanced software
- Axisymmetric Drop Shape Analysis (ADSA)
- Contact angle resolution and accuracy of 0.01° and 0.1° respectively

# AFM Analysis of Calcite Surface

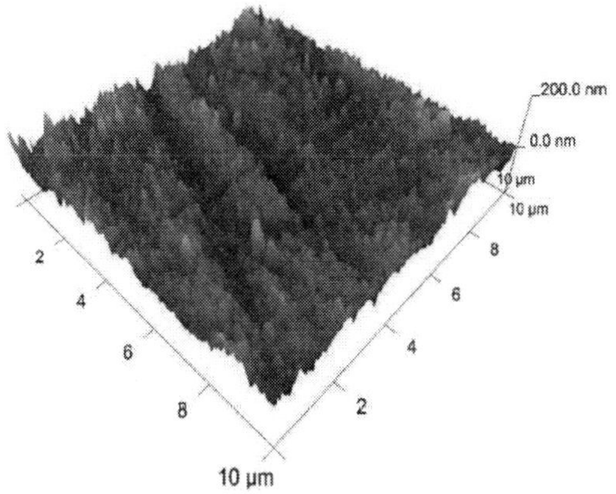


- Calcite crystals polished with 1200 grit diamond lap
- AFM analysis on 10 µm area
- Mean Roughness (Ra) - 19.9 nm

# Preliminary Experiments

- Brine concentrations
  - 0.1M NaCl and 5M NaCl
  - Calcite surface was water-wet
- Acid numbers
  - 0.3, 1.5 and 3.0
  - Acid number (AN) 1.5 is reported
- Oil Aging times
  - 1, 2 and 3 week(s)
  - 2 weeks aging is reported
- Aging temperature
  - 90 °C and 120 °C
  - 120 °C is reported

# Main Experimental Procedure

- Polishing of
  - Calcite crystal using 220, 600 and 1200 grit diamond laps
- Aging in
  - 5M brine solution for a day
  - Acid number 1.5 (AN 1.5) model oil at 120 °C for 2 weeks
  - Smart water at 120 °C for 2 weeks.
- Smart water
  - 0.164 M ionic strength (equivalent to four times diluted sea water)
  - $MgCl_2$, $MgSO_4$, $Na_2SO_4$ and NaCl solutions
- Contact angle data on
  - Oil aged and smart water aged surfaces

9

# Contact Angle Experimental Matrix

| | 1B | 1T | 2B | 2T | 3B | 3T | 4B | 4T | 5B | 5T | 6B | 6T | 7B | 7T | 8B | 8T |
|---|---|---|---|---|---|---|---|---|---|---|---|---|---|---|---|---|
| 5M Brine | X | X | X | X | X | X | X | X | X | X | X | X | X | X | X | X |
| AN 1.5 Oil | X | X | X | X | X | X | X | X | X | X | X | X | X | X | X | X |
| $MgCl_2$ | X | X | X | X | | | | | | | | | | | | |
| $MgSO_4$ | | | | | X | X | X | X | | | | | | | | |
| $Na_2SO_4$ | | | | | | | | | X | X | X | X | | | | |
| NaCl | | | | | | | | | | | | | X | X | X | X |

# Tilting Plate RCA and ACA measurement

**Surface 7B – 45 degree tilt**

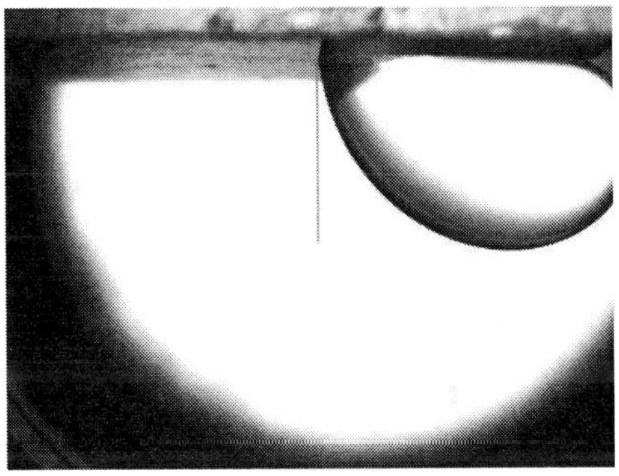

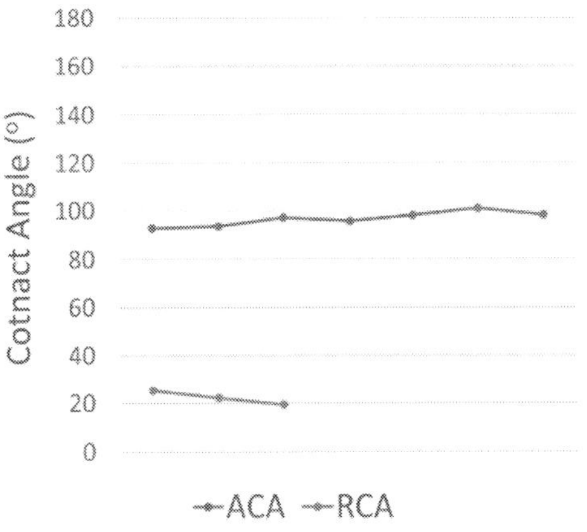


# AN1.5 Oil Aging – 14 days

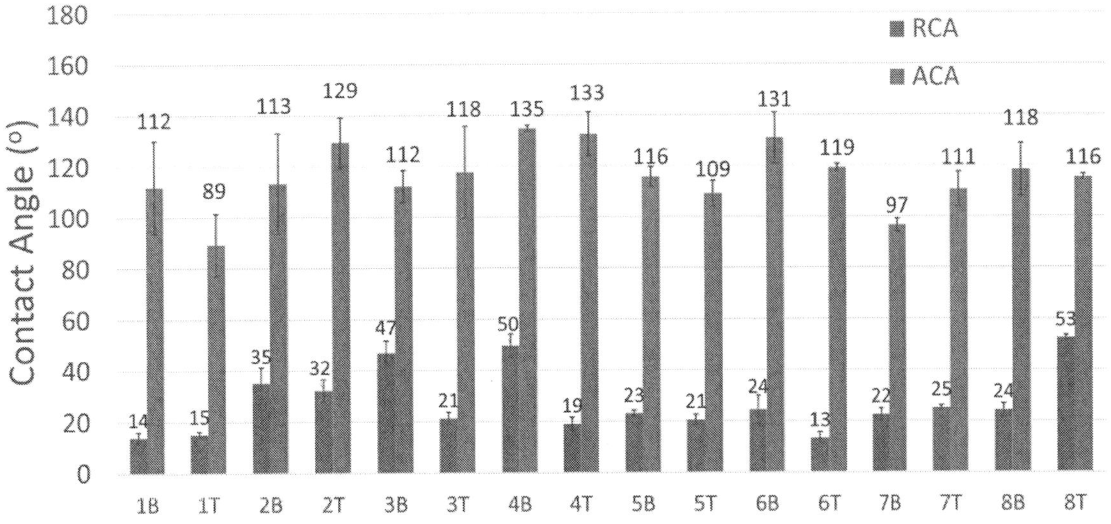

# Capillary number, $N_C$

- Capillary numbers
  - calculated from $\frac{\vartheta_{ow} \cdot \mu_w}{\sigma_{ow}}$ where
    - $\vartheta_{ow}$ - three-phase contact line velocity
    - $\mu_w$ - viscosity of the displacing (brine) fluid, 1.7 cp
    - $\sigma_{ow}$ - interfacial tension between oil and brine, 26 mN/m

| | 1B | 1T | 2B | 2T | 3B | 3T | 4B | 4T | 5B | 5T | 6B | 6T | 7B | 7T | 8B | 8T |
|---|---|---|---|---|---|---|---|---|---|---|---|---|---|---|---|---|
| $N_C \times 10^7$ | 1.57 | 2.06 | 7.67 | 0.90 | 23.28 | 0.19 | 0.55 | 0.74 | 4.00 | 4.08 | 2.78 | 14.09 | 3.38 | 2.54 | 0.50 | 0.37 |

  - ranged from $1.91 \times 10^{-8}$ to $2.33 \times 10^{-6}$
  - typical $N_C$ for flow through reservoir pores is ~ $10^{-6}$
  - <<1 denoting interfacial forces are dominating and viscous forces are negligible

# Tilting Plate RCA and ACA measurement

**Surface 8T – NaCl Aged Crystal**

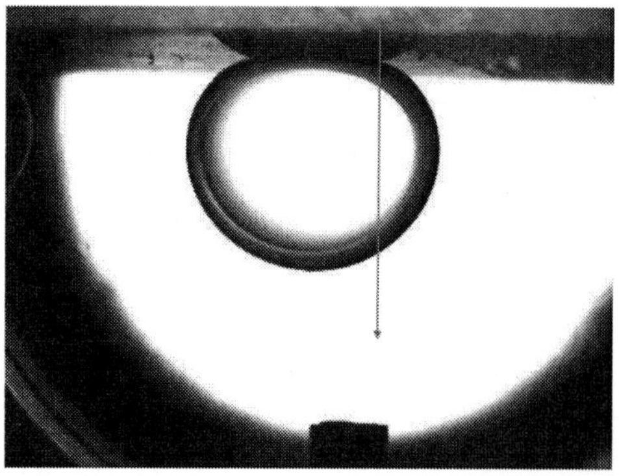

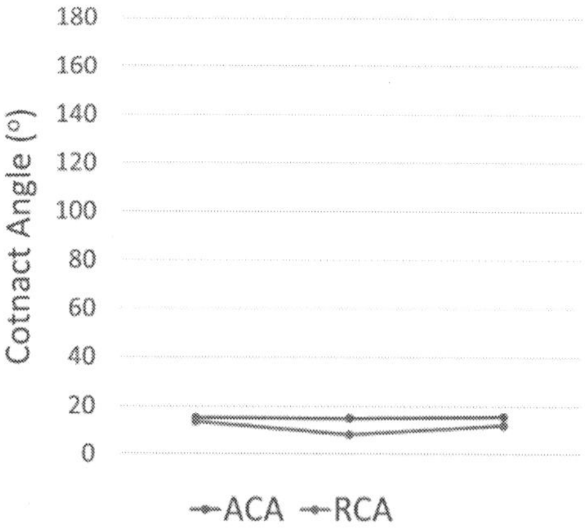

## Smart Water Aging – 14 days

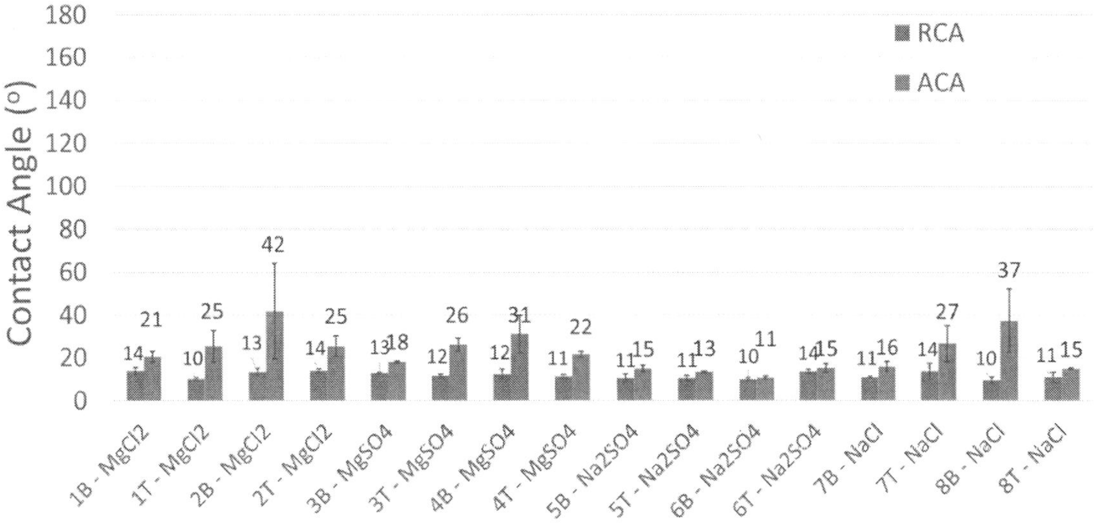


## Conclusion

- The calcite surface wettability was altered from water-wet to intermediate/oil-wet state during AN 1.5 oil aging
- Smart water aging of calcite surfaces changed the wettability back to water-wet state irrespective of the salt type
- Wettability alteration during smart water aging can be attributed to low salinity effects

Thank you!

# Assessment of Non-Linearities on the Transport Properties of Compressible Fluids

*Yuan-Yun Lin, University of Houston, USA*
*Michael T. Myers, University of Houston, USA*

## Introduction

According to the United States Energy Information Administration, natural gas production in the United States had a peak increase of 50% since 2005. Unconventional resources are expected to play an important role in future production. A significant amount of work has recently been performed to understand the role of pore structure and mineralogy on the rock properties of unconventional resource plays. [1-3]

In order to relate these observations with measurements, improved methods of measuring and interpreting plug-based measurements need to be developed. The Gas Research Institute (GRI) measurement protocol is currently the most commonly used method to measure permeability in these unconventional reservoirs. [4-9] Unfortunately, this technique requires grinding of the samples, which significantly alters the transport properties and interpretation of the data. [8-13]

The alternative discussed in this paper is to develop measurement protocols for plug scale measurements. Many earlier papers interpret plug scale transport property measurements using low - pressure drops and their associated low flow rates. This allows the linearized transport equation to be applied, but the resulting flow rates are extremely low and difficult to measure. [14-16]

This work highlights the necessity of using the full nonlinear flow equations for the interpretation of data measured at higher flow rates. We emphasize the role of gas slippage, because of its large magnitude in the permeability range that we are interested in (1-1000 nD). We limited ourselves to a 100 psi pressure differences due to the influence of large effective stress (confining stress - pore pressure) gradients on low permeability samples. Typical in-situ stresses are a few thousand psi. Applying a pressure drop of the same order of magnitude introduces too large an effective stress gradient to give meaningful data. This magnitude of pore pressure change introduces an order of magnitude permeability difference between the inlet permeability and the permeability at the outlet. Smaller pressure drops (about 10 psi) generate flow rates that are too low to be reliably measured.

We assume constant rock compressibility and fluid viscosity. Constant compressibility is consistent with moderate pore pressure changes for samples measured at higher than in situ effective stress. The stress dependent compressibility of the rock matrix $C_\phi$ (p) may be included in the modeling, but at the pressure gradients (100 psi) modeled is assumed negligible. For larger stress gradients (1000'S psi) this effect would need to be included. Viscosity is also assumed to be constant. For helium, nitrogen and methane, the viscosity varies by about one percent over the range of pressures modeled. We do not include inertial effects (Forchheimer terms[17]) due to the low flow rates encountered in typical measurements.

These assumptions lead to the one-dimensional nonlinear equation for the flow of gas in porous media:

$$\phi\mu(C_\phi + C_f)\frac{dP}{dt} = k\frac{d^2P}{dz^2} + (kC_f + \frac{dk}{dP})(\frac{dP}{dz})^2$$

Where $\phi$ is the porosity of porous media, $\mu$ is fluid viscosity, $C_\phi = 1/\phi(d\phi/dP)$ is the rock compressibility,

$C_f$ is the compressibility of the fluid, k is permeability, z is the spacial coordinate in the direction of flow and P is the pressure. The first two terms are the well-known diffusion equation. The nonlinear terms result from the compressibility of the gas and pressure dependent permeability (slippage effects). A correct handling of these nonlinear terms is essential for modeling the transport of gas in low permeability samples.

Two different measurement protocols are modeled, including:

1. Unsteady State: The upstream reference chamber is charged to an initial pressure, and the gas is allowed to flow through the sample, typically into ambient pressure at the outflow end.
2. Pulse Decay: adds an additional downstream reference chamber to the unsteady state permeability measurement. The upstream and downstream chambers are charged to different pressures, and allowed to equilibrate by flow through the sample.

## Unsteady State

### *Experiment Design*

The unsteady state permeability measurement adds an upstream reference chamber ($V_{up}$). In this model, the reference chamber is charged to an initial pressure, and the outlet pressure of the sample is assumed to be zero psi. The measurement setup is shown in Fig. 1.[16, 18]

### *Modeling Setup*

The initial conditions are:

$$p_{up}=100, \quad t=0$$
$$p_{sample}=0, \quad 0<z<L, \quad t=0$$

$P_{up}$ is the pressure in the upstream reference chamber, and is time dependent. This is the pressure that is monitored to estimate the sample transport properties. The sample pressure profile is assumed to be initially constant ($p_{sample}$). The outlet pressure ($p_{outlet}$) is zero gauge pressure:

$$p_{outlet}=0, \quad z=L, \quad t>0$$

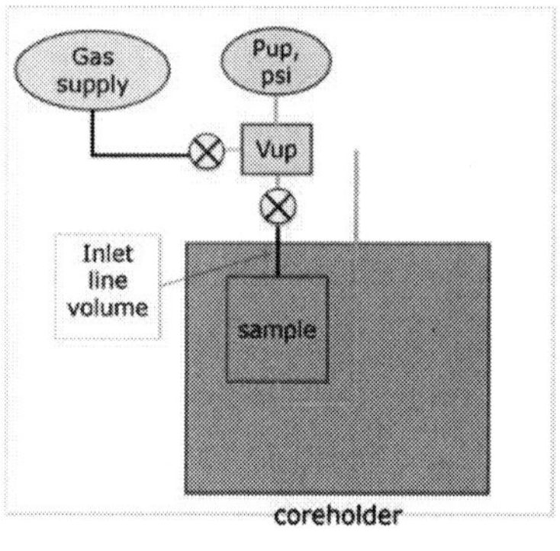


**Figure 1.** The block diagram for unsteady state permeability measurement. The outlet pressure is usually zero gauge pressure. A wide range of flow rates and pressures are involved in a single measurement.

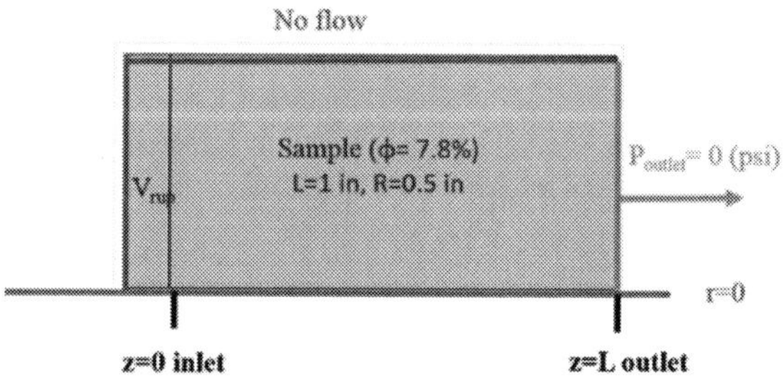


**Figure 2.** The exported COMSOL geometry used for the unsteady state modeling. The initial pressure of the upstream chamber was assumed to be 100 psi, and the initial sample pressure was assumed to be zero gauge pressure.

### *Unsteady State Modeling: No Slippage*

The time dependence of the upstream chamber pressure is shown in Fig. 3. No gas slippage was assumed (b=0 psi). For unsteady state permeability measurements the upstream chamber pressure data is all that is typically measured. The rate of decrease of the pressure is related to the mass flow rate, allowing the parameters controlling permeability to be determined ($k_0$ and b).

Fig. 4 shows the time dependent pressure profiles for the unsteady state model. The nonlinear pressure profile results from the same pressure dependent density effects exhibited in the steady state models

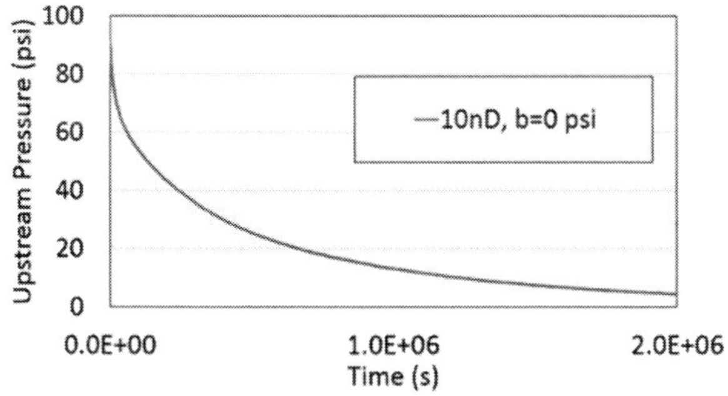


**Figure 3.** The pressure change of the upstream chamber as a function of time for the unsteady state model. The small fixed chamber size solves the problem of measuring the low flow rates inherent in steady state data.

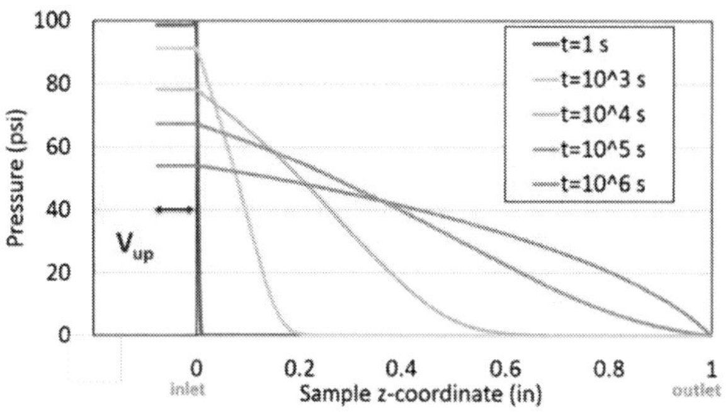


**Figure 4.** The pressure profiles for the sample and the upstream chamber for the unsteady state model (b=0 psi). The chamber pressure to the left decreases as gas flows through the sample.

### *Unsteady State Modeling: With Slippage*

The impact of the Klinkenberg correction on the unsteady state permeability is modeled in this section. As shown in Fig. 5, with increasing b the pressure declines more quickly. These models indicate that unsteady state tests only require a few days to perform with typical values of $k_0$ and b.

Fig. 6 shows the pressure profile of the upstream chamber and sample when b = 100 psi. The pressure equilibrates significantly faster when the Klinkenberg correction is added. [19]

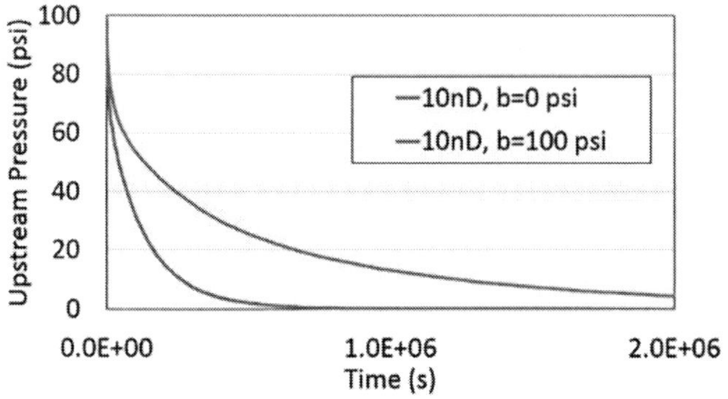


**Figure 5.** The pressure change of the upstream chamber as a function of time for the unsteady state model. With increasing b, the pressure declines more quickly. Tests for a ten nano-darcy sample typically need a few days in order to acquire sufficient data to fit for values of $k_0$ and b.

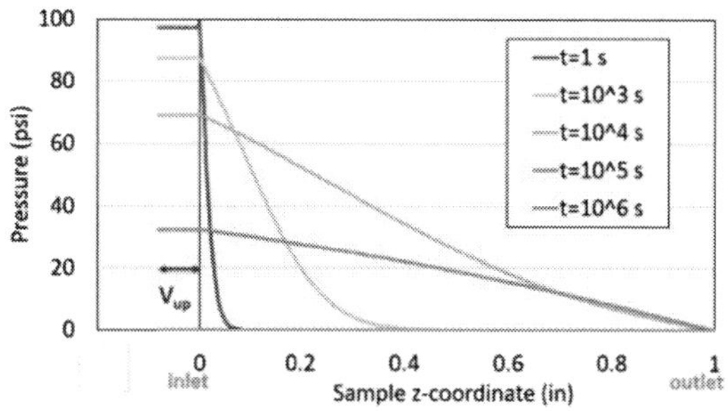


**Figure 6.** The pressure profiles for b=100 psi. Compared with Fig. 4, the pressure decline is significantly greater when slippage is included.

***Non-uniqueness of the Fit for $k_0$ and b***

Figures 7 and 8 show an example of using multiple different mean pressure measurements to determine $k_0$ and b.

Figure 7 shows four unsteady state models with initial upstream chamber pressures of 100, 200, 300 and 400 psi, while outlet pressures are 0, 100, 200 and 300 psi. These show identical fits to the chamber pressure data for a range of values of $k_0$ and b. This demonstrates the difficulty of interpreting the data for unique values of $k_0$ and b using the upstream chamber pressure alone at a single mean pressure.

Figure 8 shows the relation of $k_0$ and b for three different mean pressure models. The solutions cross at the forward model values of $k_0$=10 nD and b=100 psi.

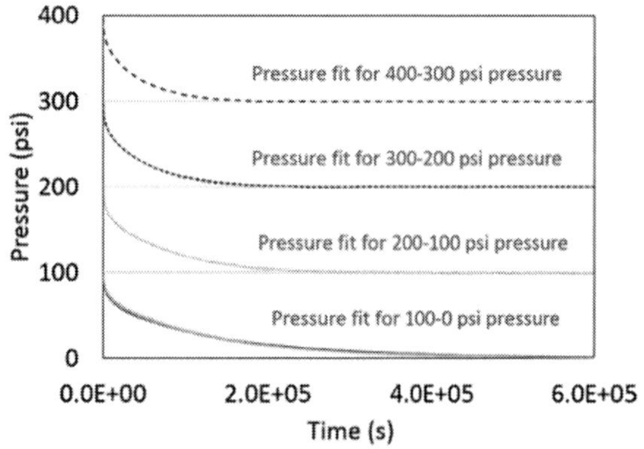


**Figure 7.** Model results for upstream chamber pressure assuming model parameters of $k_0$=10 nD and b=100 psi. The curves are an overlay of models generated by the several different sets of parameters and plotted in Fig. 8. The fits are well within experimental error.

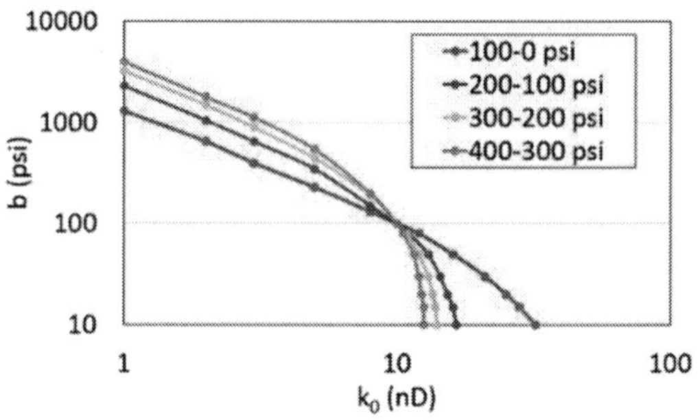


**Figure 8.** Plot of the fits for $k_0$ and b from Table 1. The lines cross at the forward modeled values for permeability and slippage. ($k_0$=10 nD and b=100 psi).

***Experimental Validation of the non-unique nature of $k_0$ and b***

Figure 9 is an example of the non-unique nature of the fit for $k_0$ and b to laboratory data. To within experimental error, the data are fit by a range of permeability ($k_0$) and gas slippage factors (b).

Figure 10 shows the relation between permeability ($k_0$) and gas slippage factor (b) in Figure. 9. Another measurement at a different mean pressure is required to separate the parameters.

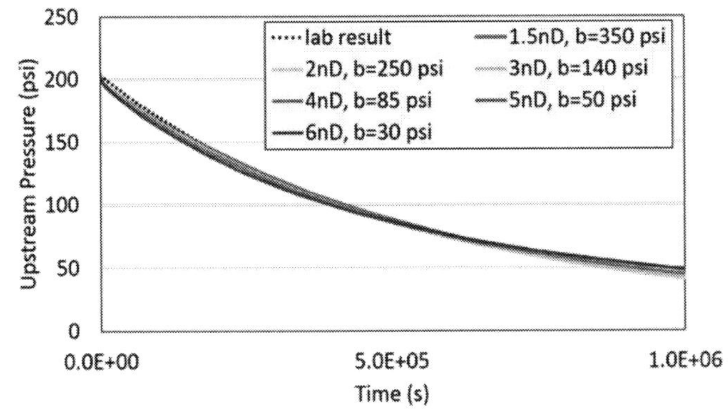


**Figure 9.** The time dependent pressure data for an unsteady state lab measurement, and fits for $k_0$ and b. All of the fits are within experimental error.

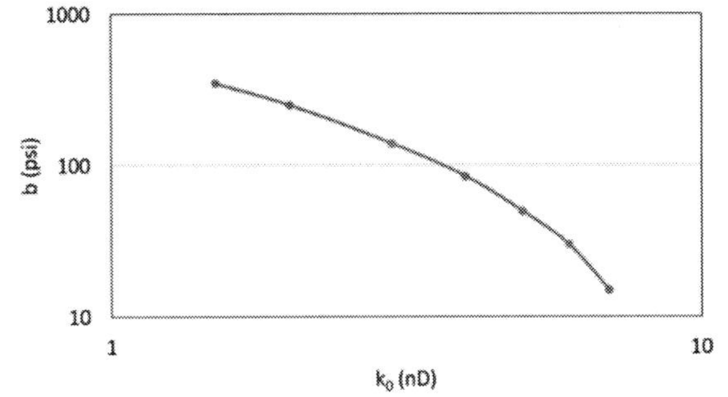


**Figure 10.** The $k_0$-b plot for the laboratory data shown in Fig. 9. As shown in Figure 7, a second measurement at a different mean pressure is necessary.

# Pulse Decay Modeling

***Experiment Design***

The pulse decay permeability measurement adds an additional downstream reference chamber ($V_{down}$). The upstream ($V_{up}$) and downstream chambers are charged to different pressures, which equilibrate by flow through the sample. The main advantage of this technique is that the system is now closed, and mass balance is maintained. Because the chamber and pore volumes are constant, sample adsorption affects the equilibrium pressure. [20-23]

***Modeling Setup***

The initial conditions are uniform pressure throughout the sample and downstream chamber:

$$p_{sample}=p_{down}=0, \quad t=0$$

$$p_{up}=100, \quad t=0$$

Where $p_{up}$ and $p_{down}$ are the upstream and downstream chamber pressures. The chamber volumes are assumed to be one cubic centimeter.

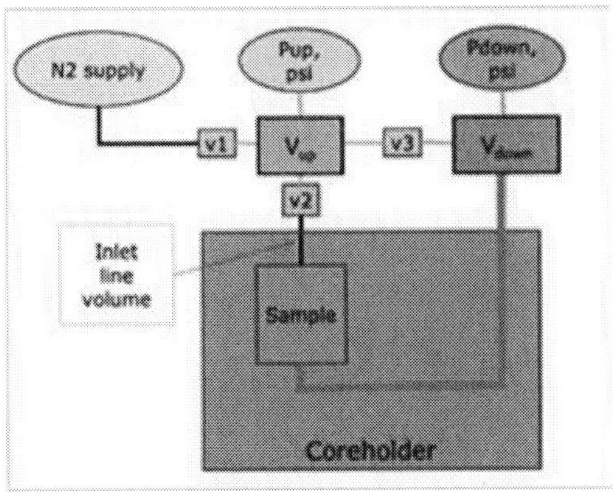


**Figure 11.** Block diagram for pulse decay permeability measurement. The addition of both upstream and downstream reference volumes allow mass balance to be maintained.

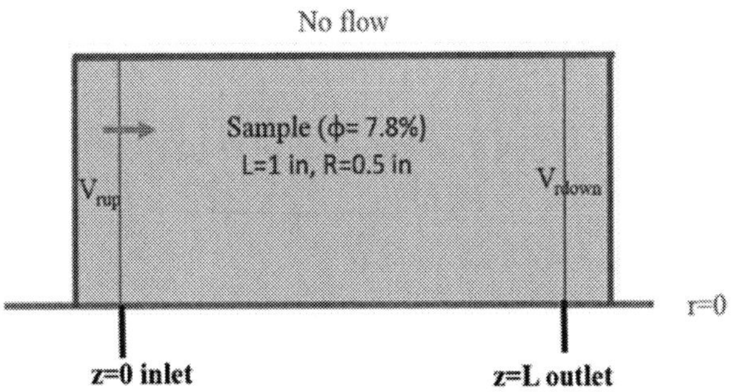


**Figure 12.** The exported COMSOL geometry used for pulse decay modeling. The initial pressure of the upstream chamber was assumed to be 100 psi. The sample and downstream chambers are assumed to be initially at zero psi.

### *Pulse Decay Modeling: No Slippage*

The time dependence of the upstream and downstream chamber pressures are shown in Fig. 13. The time delay in the downstream pressure plot is due to the finite time required for the pressure to reach the downstream chamber.

Figure 14 is the time dependent pressure profile. Similar to unsteady state, a nonlinear pressure profile is observed.

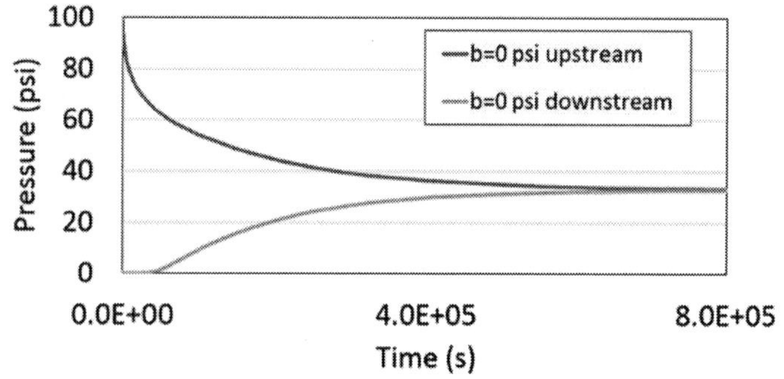


**Figure 13.** The modeled chamber pressure for the pulse decay model. The upstream chamber and pore volume are both one cubic centimeter. Consistent with the ideal gas law, the equilibrated pressure is 33.6 psi. Differences between this modeled pressure and an experimentally measured pressure are assumed to be due to adsorption effects (no leaks).

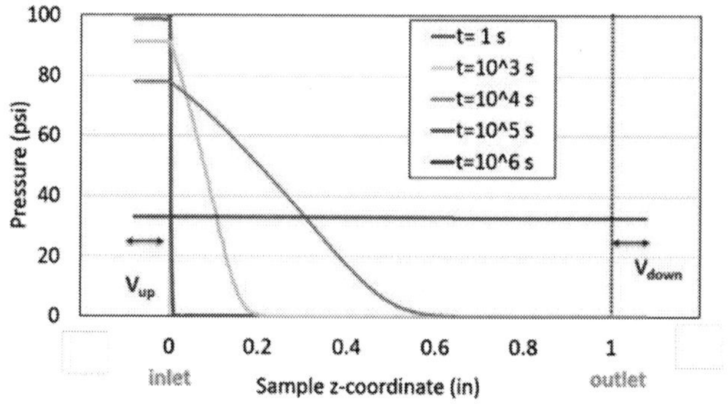


**Figure 14.** The time dependent pressure profile for the pulse decay model. The pressure at equilibrium is consistent with the idea gas law. The chamber and pore volumes are equal.

### *Pulse Decay Modeling: With Slippage*

Figure 15 includes the time dependence of the upstream and downstream chamber pressures for a Klinkenberg correction of b=100 psi. The chamber pressures equilibrate more quickly when slippage is included.

Consistent with the ideal gas law, the equilibrated pressure is 33.3 psi. Differences between this modeled pressure and an experimentally measured pressure are due to adsorption effects (assuming no leaks). Figure 16 is the time dependent pressure profile for the 10nD, b=100 psi model. The pressures equilibrate faster than for the 10nD, b=0 psi model, which is shown in Fig. 14.

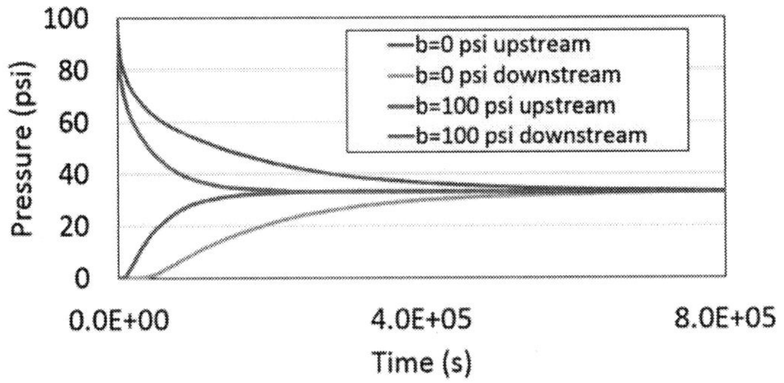


**Figure 15.** The modeled chamber pressure for the pulse decay model including gas slippage. The upstream chamber and pore volume are both one cubic centimeter. Consistent with the ideal gas law, the equilibrated pressure is 33.6 psi. Differences between this modeled pressure and an experimentally measured pressure will be assumed to be due to adsorption effects (no leaks).

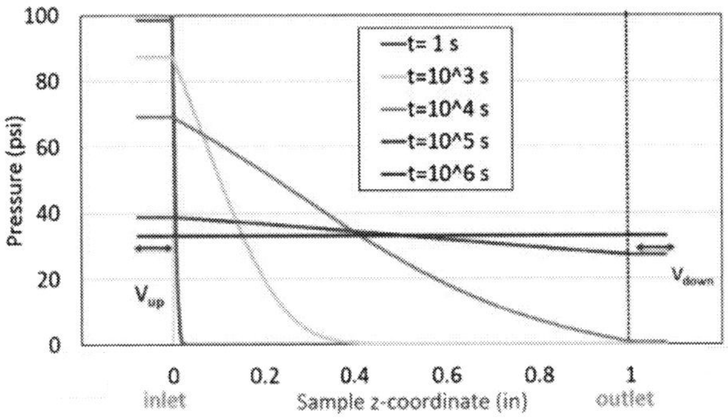


**Figure 16.** The time dependent pressure profiles for the pulse decay model including gas slippage. The chamber and pore volumes are equal. The pressure at equilibrium is consistent with the idea gas law. Comparison to Fig. 14, illustrates that the pressure equilibrates more quickly.

## *Non-uniqueness of the fits for $k_0$ and b*

Figure 17 shows the time dependent pressures in the upstream and downstream chambers for different mean pressures. The initial upstream pressures are 100, 200, 300 and 400 psi while the initial downstream chamber pressures are 0, 100, 200 and 300 psi, respectively.

Figure 18 shows the relation of $k_0$ and b for four different mean pressures. A wide range of permeabilities and Klinkenberg b values fit the data. They cross at the forward modeled values.

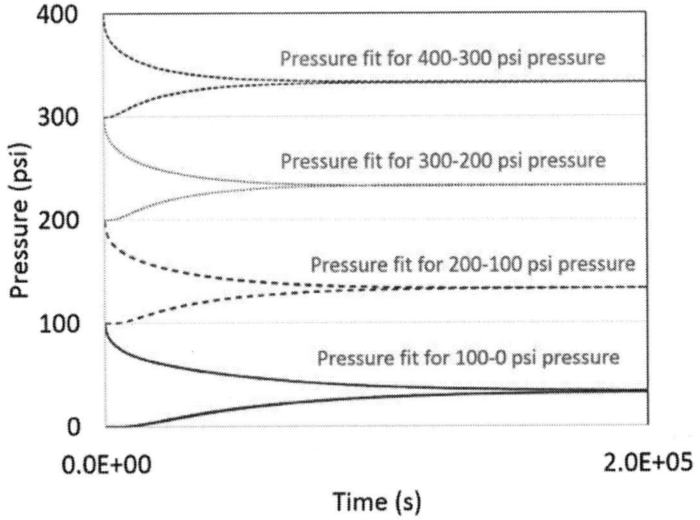


**Figure 17.** Forward Modeled data for $k_0$=10 nD and b=100 psi. The curves are an overlay of models generated by the several sets of parameters which are plotted in Fig. 18. The fits are well within experimental error.

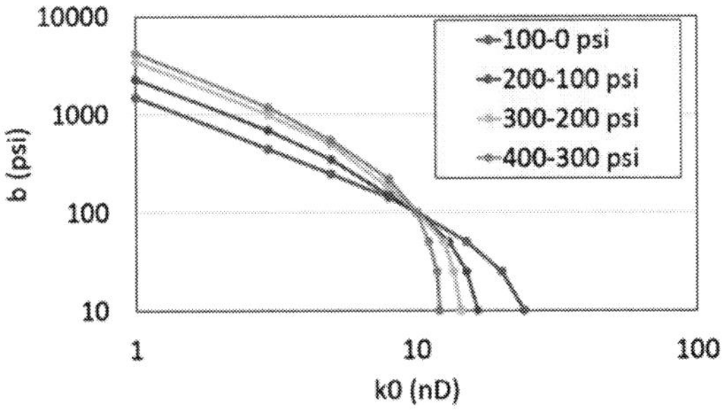


**Figure 18.** $k_0$-b plot for the COMSOL pulse decay model from Fig. 17. They cross at the forward modeled values of $k_0$=10 nD and b=100 psi.

*Experimental Validation of the non-unique nature of $k_0$ and b*

Figure 19 is an example of the non-unique nature of the fit for $k_0$ and b to laboratory data for experimental pulse decay data. To within experimental error the data are fit by a range of permeability ($k_0$) and gas slippage factors (b).

The reason for the mismatch between the models and measured data may be due to gas absorption, lack of early time data in the pressure curves or errors in the porosity measurement. The source of these errors will be the object of future investigation.

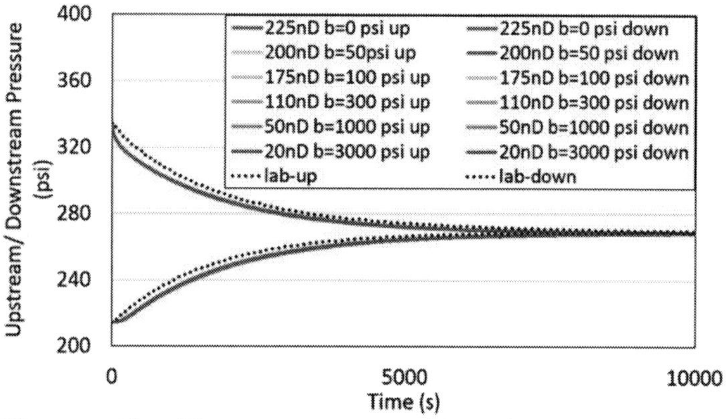


**Figure 19.** The pressure profile of a pulse decay lab measurement, and fits for $k_0$ and b. All of the fits do not allow the values of to be determined from a single measurement. Measurements at two different mean pressures are required.

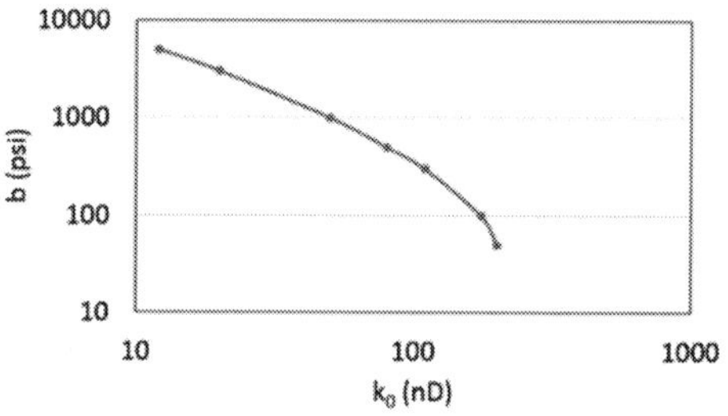


**Figure 20.** The $k_0$-b plot for the laboratory data shown in Fig. 19. The uncertainty in the parameters are large making a second measurement at a different mean pressure necessary.

Figure 20 shows the relation between permeability ($k_0$) and gas slippage factor (b) in Figure 19. Another measurement at a different mean pressure is required to separate the parameters.

Each of the data fitting are within RMS error for 7.2 %.

## Conclusions

We have developed models for a variety of geometries and experimental protocols:

1. Unsteady State: This technique takes the least time to perform the measurement over a wide range of flow rates. At low permeability, the full nonlinear equation must be used to extract the correct value of $k_0$ and b. Measurements at two or more mean stresses are required to allow unambiguous extraction of the transport parameters. The measurement time is comparable to many other petro-physical measurements (days). This technique should become the standard for low permeability plug measurements.

2. Pulse Decay: The measurement provides mass balance, which allows the effects of absorption and adsorption to be included in models. This makes it the most robust measurement for determining all the physical mechanisms, but small leaks would negate these advantages. Similar to unsteady state measurements at several means stresses should be made.

The modeling indicates that plug scale measurements are practical, but multiple mean pressures must be used to separate the competing effects of permeability and gas slippage. Typical measurement times are on the order of days and grinding of the samples used for the standard GRI technique is therefore not necessary. We recommend using unsteady state measurements at reservoir stress, supplemented with pulse decay to calibrate the nonlinear effects, i.e. the impact of diffusion, adsorption and absorption.

***References***


1. Hathon, L.A. and M.T. Myers, *Shale Rock Properties: Peak Strength, Acoustic Anisotropy And Rock Fabric*. 2012, American Rock Mechanics Association.
2. Hooghan, K., M. Dixon, and L. Hathon, *Imaging Shales: How low can you go? Why?* 2014: SPE Nanotechnology and Nano-Geoscience in Oil and Gas Industry Workshop.
3. Hooghan, K., M. Dixon, and L. Hathon, *Correlating Core to Pore locations and Comparing with Rock Property Measurements: an Initial Foray*. 2016: Weatherford Laboratories.
4. Luffel, D.L. and F.K. Guidry, *New Core Analysis Methods for Measuring Reservoir Rock Properties of Devonian Shale*. 1992.
5. Luffel, D.L., C.W. Hopkins, and P.D. Schettler, Jr., *Matrix Permeability Measurement of Gas Productive Shales*. 1993, Society of Petroleum Engineers.
6. Luffel, D.L., F.K. Guidry, and J.B. Curtis, *Evaluation of Devonian Shale With New Core and Log Analysis Methods*. 1992.
7. ResTech, H. and I. Gas Research, *Development of Laboratory and Petrophysical Techniques for Evaluating Shale Reservoirs: Final Technical Report : October 1986-September 1993*. 1996: Gas Research Institute.
8. Profice, S., et al., *Permeability, Porosity and Klinkenberg Coefficien Determination on Crushed Porous Media*. 2012.
9. Cui, X., A.M.M. Bustin, and R.M. Bustin, *Measurements of gas permeability and diffusivity of tight reservoir rocks: different approaches and their applications*. Geofluids, 2009. **9**(3): p. 208-223.
10. Tinni, A., et al., *Shale Permeability Measurements on Plugs and Crushed Samples*. 2012, Society of Petroleum Engineers.
11. Civan, F. and D. Devegowda, *Comparison of Shale Permeability to Gas Determined by Pressure-Pulse Transmission Testing of Core Plugs and Crushed Samples*. 2015, Unconventional Resources Technology Conference.
12. Mokhtari, M. and A.N. Tutuncu, *Characterization of anisotropy in the permeability of organic-rich shales*. Journal of Petroleum Science and Engineering, 2015. **133**: p. 496-506.
13. Cui, X.A., et al. *A new method to simultaneously measure in-situ permeability and porosity under reservoir conditions: Implications for characterization of unconventional gas reservoirs*. in *Canadian Unconventional Resources and International Petroleum Conference*. 2010. Society of Petroleum Engineers.
14. Lin, W., *Compressible fluid flow through rocks of variable permeability*. 1977.
15. Fischer, G.J., *Chapter 8 The Determination of Permeability and Storage Capacity: Pore Pressure Oscillation Method*, in *International Geophysics*, E. Brian and W. Teng-fong, Editors. 1992, Academic Press. p. 187-211.
16. Aronofsky, J.S., *Effect of Gas Slip on Unsteady Flow of Gas Through Porous Media*. Journal of Applied Physics, 1954. **25**(1): p. 48-53.
17. Forchheimer, P., *Wasserbewegung durch boden*. Z. Ver. Deutsch. Ing, 1901. **45**(1782): p. 1788.
18. Jones, S.C., *A Rapid Accurate Unsteady-State Klinkenberg Permeameter*. 1972.

19. Rushing, J.A., et al., *Klinkenerg-Corrected Permeability Measurements in Tight Gas Sands: Steady-State Versus Unsteady-State Techniques*. 2004, Society of Petroleum Engineers.
20. Bourbie, T. and J. Walls, *Pulse Decay Permeability: Analytical Solution and Experimental Test.* 1982.
21. Brace, W.F., J.B. Walsh, and W.T. Frangos, *Permeability of granite under high pressure.* Journal of Geophysical Research, 1968. **73**(6): p. 2225-2236.
22. Dicker, A.I. and R.M. Smits, *A Practical Approach for Determining Permeability From Laboratory Pressure-Pulse Decay Measurements*. 1988, Society of Petroleum Engineers.
23. Jones, F.O. and W.W. Owens, *A Laboratory Study of Low-Permeability Gas Sands.* 1980.

# 509054 Solve this! Fundamental Approach to Problem Solving in Industrial Processes I (Invited Talks)

**Abstract:**

Solving complex plant problems is an everyday exercise that affects all projects and budgets, and it is very rarely introduced to students at any level. In fact, by the time young engineers are few years into working in industry, most have lost the engineering skills that they developed in school. We hope to encourage future engineers to retain those skills and learn the importance of plant problem-solving.

# 497067 Modeling Adsorption and Transport Processes in Kerogen

**Abstract:**

A better elucidation of adsorption and transport processes in chemically heterogeneous and complex geomaterials such as kerogen can play a significant role in an improved understanding of phenomena such as retention of hydrocarbons and small fluid molecules (e.g., $CO_2$) in kerogen and their expulsion from it. In the present investigation, a Grand Canonical Monte Carlo/Molecular Dynamics simulation approach is used to model adsorption and transport of pure $CH_4$ and $CO_2$ in a type II oil window kerogen, and subsequently to determine the adsorption and transport properties of $CH_4$ and $CO_2$ at various reservoir conditions. The $CO_2$ exhibits relatively higher adsorption in type II oil window kerogen than $CH_4$ at given reservoir temperature and pressure. Further, the adsorption capacity of both $CO_2$ and $CH_4$ decreases with an increase in reservoir temperature. Finally, the transport of $CH_4$ and $CO_2$ molecules within kerogen matrix is highly influenced by the fluid-surface interaction and availability of the transport pathways within the kerogen matrix.

# 499607 Impact of Hydraulic Fracture Permeability on Oil Production

**Abstract:**

Oil and gas production from tight formations or shales has become possible due to technological advancements in horizontal drilling and hydraulic fracturing. Proppants such as sand mixed with fracturing fluids are pumped at high pressures to crack the reservoir formation rock creating hydraulic fractures. The effectiveness of a fracturing job is measured by the dimensions of the fracture and ease of hydrocarbon flow within the fracture, i.e. the fracture permeability. Since hydraulic fractures connect the wellbore to the formation, economic production for any well is highly dependent on how conductive these fractures are. The objective of this study is to investigate the impact of hydraulic fracture permeability on oil and gas production performance in terms of fluid rate and recovery. Five reservoir fluids (two volatile oils, two gas-condensates and one dry gas case) are tested in a 100 nDpermeability reservoir with two operating bottomhole pressures of 1000 psia and 3000 psia. The flow rates and recovery for most fluids increase with fracture permeability up to 1000 mD. We found that fractures with a permeability above 1000 mD act as infinitely conductive to fluids leading to no further improvements in rates and recovery. However, recovery is not always proportional to fracture permeability. For the cases of volatile oils and condensates, a crossover in recovery is observed where the production is higher for low fracture permeability compared to higher permeability fractures. Rates are very high for higher fracture permeability cases, but rates drop sharply (80% of initial rate in 1 year). Rates for lower fracture permeability cases are not as high, but do not decline as sharply either. In some instances, lower fracture permeabilities outperform higher fracture permeability cases in terms of cumulative production. Although higher fracture permeability is essential for high flow rates, higher pressure drop near fractures causes gas to come out of oil solution, eventually dominating the multiphase flow and suppressing liquid rate. A similar phenomenon also happens for volatile and condensate fluids. An optimal fracture permeability is necessary to extract the maximum amount of oil or gas. The type of fluid, operating bottom hole pressure and reservoir permeability are the main factors to determine the optimum fracture permeability.

# 511762 Fundamentals of Wave Propagation and Development of a Control Strategy for Pump Control in Hydraulic Fracturing

**Abstract:**

A stimulation job entails high pressure pumping of fluid and proppant into a well-bore to induce hydraulic fractures in the subsurface formation, typically along a horizontal well. There may be 10-25 reciprocating pumps in simultaneous operation. The pressure pulses from the pumps induce wave-guided acoustic modes in the pipes that travel at the wave speed along the pipe. When these bounce off a reflecting surface (such as a valve or a bend in the pipe) they generate standing waves that may combine constructively or destructively. When the piping system comprises elbows, tees, or diameter changes, pressure pulsations can lead to piping vibrations.

As a reduction in the peak-to-peak pressure pulsations will lead to a concomitant reduction in vibration, an active control strategy is presented that reduces the pulsations from the reciprocating pumps by operating in a regime where the stationary waves combine destructively at a majority of points in the piping system.

A 10-pump Flowmaster® CFD simulation model was developed and calibrated using experimental data obtained from a physical 2-pump system. This model was incorporated in an optimization framework designed to minimize the peak-to-peak (P2P) pressure differential observable at multiple locations with active pump speed and phase control. The procedure commences from a given starting condition, the prevailing state of the pump phases and speeds. This solution is then perturbed within specified limits over a number of consecutive steps and the variations are recorded. The process serves to qualify the lower and upper bounds, and the distribution properties, of the P2P merit function in the *Sampling Mode*, and then aims to detect good configurations yielding low P2P values in the *Control Mode*. When a solution within the designated tolerance of the lower bound estimate is found, the incumbent design configuration (pump phases and speeds is retained as the active solution). The process reverts to the *Sampling Mode* if the conditions change appreciably. Results for implementation of this control method will be presented. A solution close to the estimated lower bound is established, that is considerably far from the upper bound. This demonstrates a robust and active control strategy, one that can manage any number of pumps and indeed, is applicable for active real-time control in the field, with its added complexity.

# 507046 Schlumberger's Automated Stimulation Delivery Platform

**Abstract:**

Schlumberger's next-generation automated stimulation delivery platform is a bottom-up redesign of wellsite hydraulic fracturing surface equipment and automation that brings highly industrialized and sustainable processes to a mobile operation. Automating stimulation delivery redefines industry expectations for hydraulic fracturing process control and consistency. Conventional stimulation equipment automates only a few mixing and pumping functions, which cannot meet the demands of modern hydraulic fracturing operations: enormous volumes of fluid and proppant; complex mixing requirements; and continuous, high-intensity pumping operations. The new processes with automation bring a high level of work synergies, an increase in efficiency, a 10-fold increase in reliability, and a minimized environmental footprint to the industry's growing requirements for more stages and proppant delivered per day. This multiyear effort to revolutionize the service delivery platform shows Schlumberger's commitment to innovation and global stewardship in sustainability.

This presentation will focus on the new technologies and methods that have enabled the automated stimulation delivery platform. The product development needs such as extreme engineering, synergistic collaborations, and rapid prototyping techniques will be discussed to provide an insight into how industry adapts and innovates new products in a critical time-to-market project. Furthermore, a sustainability modeling software called Stewardship Tool, developed by Schlumberger, will be utilized to demonstrate the superior environmental footprint performance of the new platform as compared to the traditional methods used in the industry.

# 496735 Potential Use of Solid Nanoparticles to Mitigate Hydrate Formation in Water-in-Oil Emulsions in the Presence of Wax

**Abstract:**

Clathrate hydrates are non-stoichiometric, ice-like crystalline solids that can lead to plugging of crude oil pipelines. In addition, wax deposition leads to partial or complete blockage of crude oil pipelines. Crude oil is a complex hydrocarbon mixture that includes asphaltenes, aromatics, naphthenes, resins, and paraffins. De-convoluting the effects of solid particles and surfactants that are present in crude oil on hydrate and wax formation would improve the hydrate and wax prediction strategies. The effect of hydrophobicity of solid particles, surfactants, and water on hydrate and wax formation in water-in-oil emulsions was investigated. Cyclopentane hydrates are studied in model oil systems using surfactant and solid particles, which act as stabilizing agents. An Olympus BX53 polarized optical microscope with shear cell and temperature control (-50°C to 450°C) stage is used to quantify droplet size distribution, and hydrate crystals morphology. A DHR-3 stress controlled rheometer whose temperature can be controlled between -20Â °C to 150Â°C is used to examine the rheological behavior of hydrate forming emulsions. Furthermore, a differential scanning calorimeter (DSC) was used to determine (i) the emulsion stability, (ii) the extent of hydrate conversion, and (iii) wax appearance temperature.

The results showed that the presence of silica and water did not have a significant effect on the wax appearance temperature (WAT). In addition, the amount of hydrate formation decreased with an increase in hydrophobicity of silica nanoparticles at the waterâˆ'oil interface. The presence of wax promoted hydrate formation in water-in-oil emulsions stabilized using either highly hydrophobic silica nanoparticles or a surfactant. Our data indicate that wax did not exist at the water/oil interface; therefore, wax did not serve as nucleation sites for hydrate formation. The addition of wax affected the droplet size of water-in-oil emulsions stabilized by highly hydrophobic silica nanoparticles, which, in turn, influenced hydrate formation. Also, hydrate dissociation led to destabilization of water-in-oil emulsions in the presence of wax that led to changes in the WAT. Rheological investigations were also carried out to determine the flow behavior and yield stress of hydrates in the presence of solid particles of different wettabilities in the presence of wax. Rheological Investigations also were carried out to examine the potential use of hydrophobic silica nanoparticles in hydrate mitigation strategies in surfactant stabilized water-in-oil emulsions.

# CORROSION MODELING USING ELECTROCHEMISTRY AND COMPUTATIONAL FLUID DYNAMICS

Electrochemistry coupled with CFD for solving mass transport and surface chemistry simultaneously

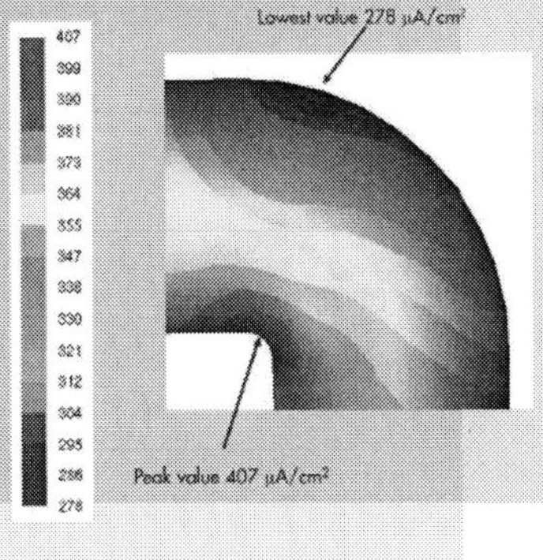


Kuochen Tsai
Shell International Exploration and Production, Inc

2017 AIChE Annual Meeting, Minneapolis, MN, Oct. 29 – Nov. 3



# DEFINITIONS & CAUTIONARY NOTE

Reserves: Our use of the term "reserves" in this presentation means SEC proved oil and gas reserves.

Resources: Our use of the term "resources" in this presentation includes quantities of oil and gas not yet classified as SEC proved oil and gas reserves. Resources are consistent with the Society of Petroleum Engineers 2P and 2C definitions.

Organic: Our use of the term Organic includes SEC proved oil and gas reserves excluding changes resulting from acquisitions, divestments and year-average pricing impact.

Shales: Our use of the term 'shales' refers to tight, shale and coal bed methane oil and gas acreage.

The companies in which Royal Dutch Shell plc directly and indirectly owns investments are separate entities. In this document "Shell", "Shell group" and "Royal Dutch Shell" are sometimes used for convenience where references are made to Royal Dutch Shell plc and its subsidiaries in general. Likewise, the words "we", "us" and "our" are also used to refer to subsidiaries in general or to those who work for them. These expressions are also used where no useful purpose is served by identifying the particular company or companies. "Subsidiaries", "Shell subsidiaries" and "Shell companies" as used in this document refer to companies over which Royal Dutch Shell plc either directly or indirectly has control. Companies over which Shell has joint control are generally referred to as "joint ventures" and companies over which Shell has significant influence but neither control nor joint control are referred to as "associates". The term "Shell interest" is used for convenience to indicate the direct and/or indirect ownership interest held by Shell in a venture, partnership or company, after exclusion of all third-party interest.

This presentation contains forward-looking statements concerning the financial condition, results of operations and businesses of Royal Dutch Shell. All statements other than statements of historical fact are, or may be deemed to be, forward-looking statements. Forward-looking statements are statements of future expectations that are based on management's current expectations and assumptions and involve known and unknown risks and uncertainties that could cause actual results, performance or events to differ materially from those expressed or implied in these statements. Forward-looking statements include, among other things, statements concerning the potential exposure of Royal Dutch Shell to market risks and statements expressing management's expectations, beliefs, estimates, forecasts, projections and assumptions. These forward-looking statements are identified by their use of terms and phrases such as "anticipate", "believe", "could", "estimate", "expect", "intend", "may", "plan", "objectives", "outlook", "probably", "project", "will", "seek", "target", "risks", "goals", "should" and similar terms and phrases. There are a number of factors that could affect the future operations of Royal Dutch Shell and could cause those results to differ materially from those expressed in the forward-looking statements included in this presentation, including (without limitation): (a) price fluctuations in crude oil and natural gas; (b) changes in demand for Shell's products; (c) currency fluctuations; (d) drilling and production results; (e) reserves estimates; (f) loss of market share and industry competition; (g) environmental and physical risks; (h) risks associated with the identification of suitable potential acquisition properties and targets, and successful negotiation and completion of such transactions; (i) the risk of doing business in developing countries and countries subject to international sanctions; (j) legislative, fiscal and regulatory developments including potential litigation and regulatory measures as a result of climate changes; (k) economic and financial market conditions in various countries and regions; (l) political risks, including the risks of expropriation and renegotiation of the terms of contracts with governmental entities, delays or advancements in the approval of projects and delays in the reimbursement for shared costs; and (m) changes in trading conditions. All forward-looking statements contained in this presentation are expressly qualified in their entirety by the cautionary statements contained or referred to in this section. Readers should not place undue reliance on forward-looking statements. Additional factors that may affect future results are contained in Royal Dutch Shell's 20-F for the year ended 31 December, 2015 (available at www.shell.com/investor and www.sec.gov ). These factors also should be considered by the reader. Each forward-looking statement speaks only as of the date of this presentation, 14 November 2016. Neither Royal Dutch Shell nor any of its subsidiaries undertake any obligation to publicly update or revise any forward-looking statement as a result of new information, future events or other information. In light of these risks, results could differ materially from those stated, implied or inferred from the forward-looking statements contained in this presentation. There can be no assurance that dividend payments will match or exceed those set out in this presentation in the future, or that they will be made at all.

We use certain terms in this presentation, such as discovery potential, that the United States Securities and Exchange Commission (SEC) guidelines strictly prohibit us from including in filings with the SEC. U.S. Investors are urged to consider closely the disclosure in our Form 20-F, File No 1-32575, available on the SEC website www.sec.gov. You can also obtain this form from the SEC by calling 1-800-SEC-0330.

## INTRODUCTIONS

- Mixing and transport of chemical species can play a dominant role in metallic corrosion process
- Recent advancement in CFD software enables surface corrosion chemistry to be incorporated with mass transport.
- Notable chemistry for oil and gas industry: $CO_2$, $H_2S$ and $O_2$
- $CO_2$ corrosion has been widely studied and its detailed chemistry and limitations in mass transport are relatively better understood.
- A combination of mass transport and corrosion chemistry can provide a more reliable predictive tool for flow assisted corrosion.



## CORROSION CHEMISTRY*

- Solution Chemistry – Carbonic acid equilibrium

$CO_2 + H_2O \Leftrightarrow H_2CO_3$

$H_2CO_3 \Leftrightarrow HCO_3^- + H^+$

$HCO_3^- \Leftrightarrow H^+ + CO_3^{2-}$

- Electro-Chemistry
  - Anodic: $Fe \rightarrow Fe^{2+} + 2e^-$ (steel dependent, Tafel slope parameters must be measured)
  - Cathodic: $2H^+ + 2e^- \rightarrow H_2$;
  - $H_2CO_3 + e^- \rightarrow H + HCO_3^-$ (if 4<pH<6)
- Exchange current density:

$i_a = i_0\left(10^{\frac{E-E_{eq}}{b_a}} - 10^{\frac{E-E_{eq}}{b_c}}\right)$ Butler-Volmer equation

$E_{eq}$- Solved by Nernst Equation ($E_{eq} = \frac{-RT}{nF}\ln(K) = -2.303\mathrm{RT}\frac{pH}{F}$)

$E$ – Overpotential (*0* for corrosion, + value for electrolytic decomposition)

- Mixed potential theory: Anodic current density = cathodic current density -> solves overpotential

*Nesic et al., Corrosion Science **52**, pp. 280, 1996

# MASS TRANSFER LIMITED CORROSION RATES

- Mass-transfer limited corrosion rate modeled by empirical Sherwood number:

$$\frac{1}{i_{(H^+)}} = \frac{1}{i_{\alpha(H^+)}} + \frac{1}{i^d_{lim}(H^+)}$$

$$i^d_{lim}(H^+) = k_m F \times [H^+]_b$$

$$Sh = \frac{k_m l}{D} = 0.15 \times Re^{0.86} \times Sc^{0.33}$$

$$i_{0(H^+)} \propto [H^+]^{0.5}$$

$$i_{o(H_2CO_3)} \propto [H_2CO_3][H^+]^{-0.5}$$

$$i^r_{lim}[H_2CO_3] = F \times [CO_2]_b \times \left(D_{H_2CO_3} K_{hyd} k^f_{hyd}\right)^{0.5}$$ - $CO_2$ hydration limited rate

- Re number is determined using local velocity ( $\sqrt{\tau_w}$ ) values near wall.
- Other mass transfer parameters possible, but corrosion rate correlates to $\sqrt{\tau_w}$ best.



# PIPE CORROSION DATA IN A STRAIGHT PIPE*

- $CO_2$ corrosion experiment in pipes with adjustable pH and velocity
- Pipe ID:15 mm
- Velocity range: 0-13 m/s
- pH range: 4 - 6

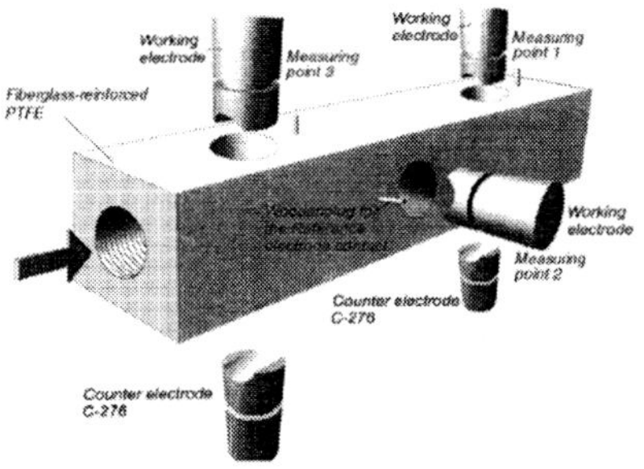


Experiment setup schematic*

*Nesic S. and Solvi G. T., Corrosion Sci. 51, pp. 773-787

# RESULTS AND DISCUSSIONS

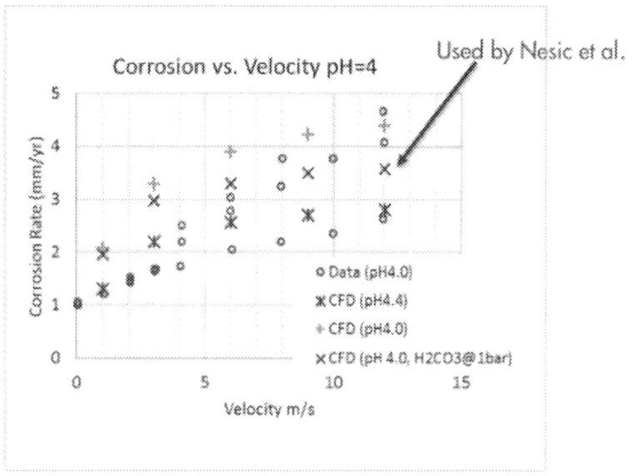


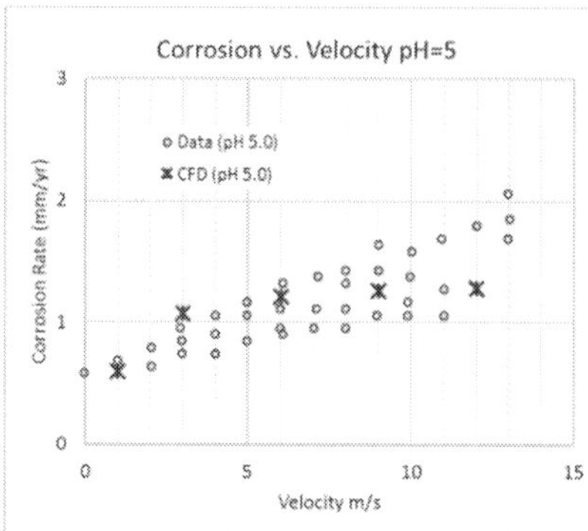


- Data scatter is significant.
- CFD model can be made to fit the data only by varying pH value, but the trend is captured well.
- At pH=5, the predicted values are within the experimental data range
- The corrosion rate are well captured at low velocity (diffusion limited)



# EXPERIMENTAL DATA FOR PIPE BEND

- Experimental data by Zhang et al*. using $CO_2$ saturated solutions with pH~6
- Data extracted from the model does not predict the measured corrosion value, a pH of 4 is needed to produce the measured corrosion peak value.
- The solution is assumed to be in equilibrium concentrations.
- Only surface chemistry is considered and species are transported by the flow.

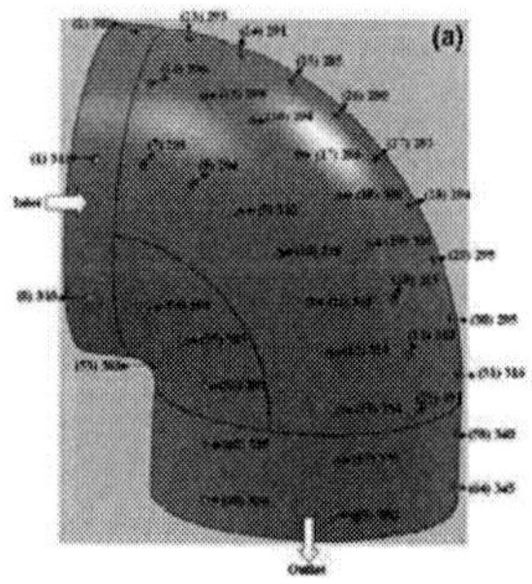


*Zhang et al., Corrosion Science 77 (2013)

# RESULTS AND DISCUSSIONS

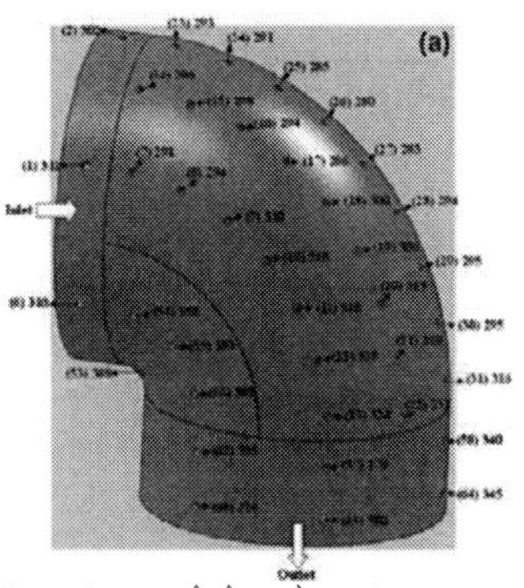


Experimental data locations

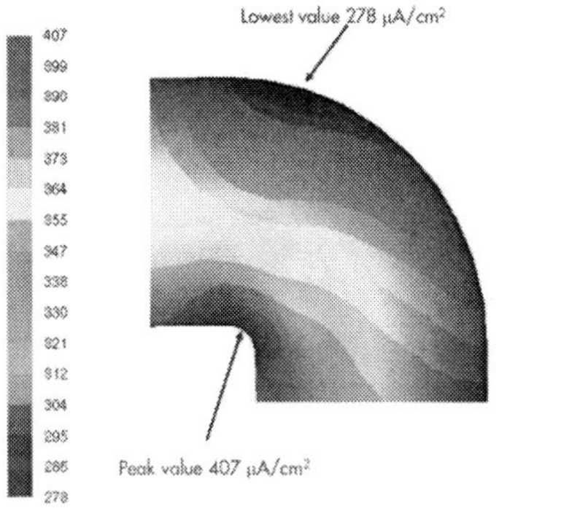


CFD predicted corrosion current density

- CFD models:
  - Realizable k-e with scalable wall function -> to obtain consistent wall shear stress values
  - $H^+$ must be present to initialize solutions.



# RESULTS AND DISCUSSIONS – CONT'D

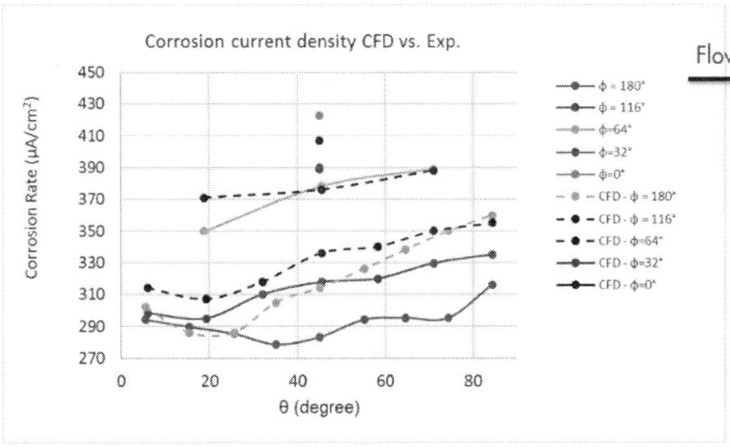


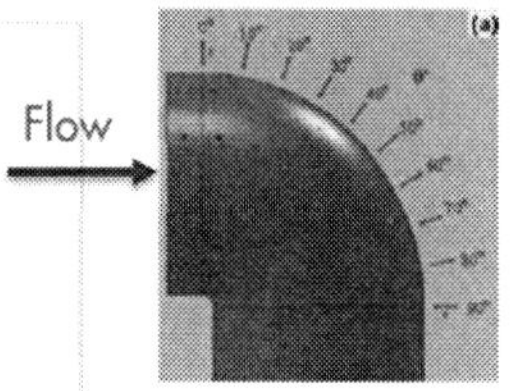


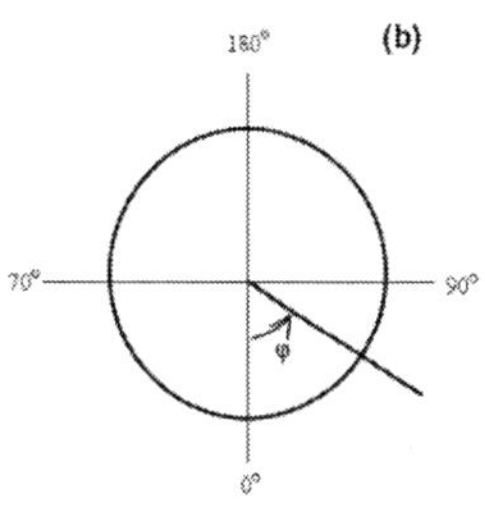


- Top values match well.
- In general the trend is well predicted

## RESULTS AND DISCUSSIONS – CONT'D

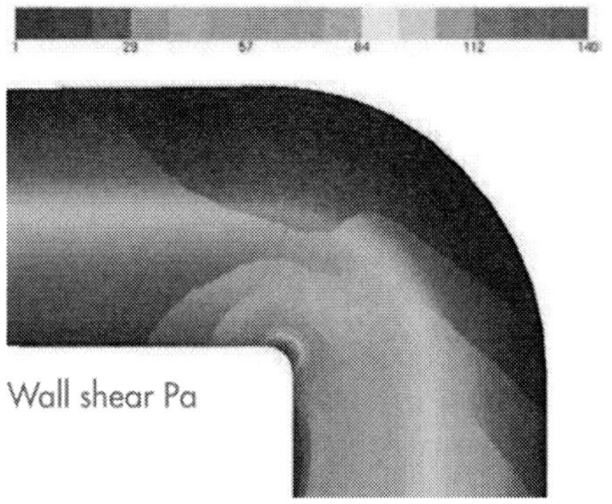


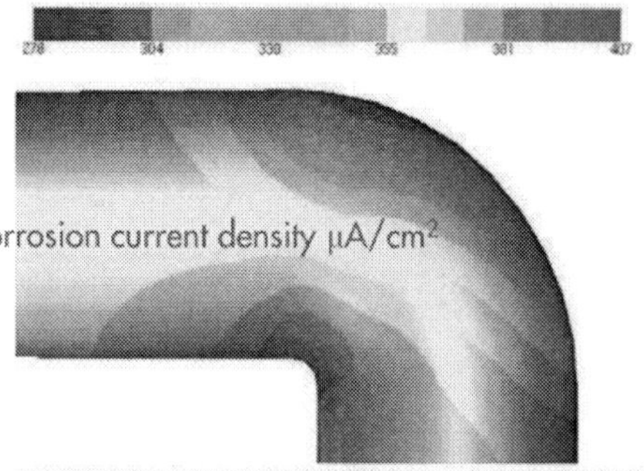


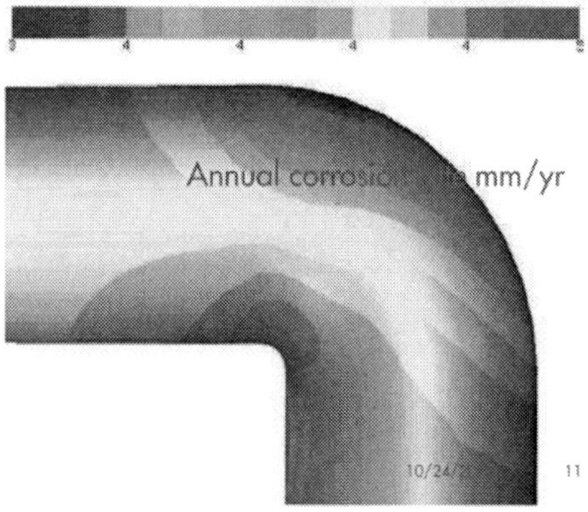


- Model can directly shows corrosion variations on surfaces and its correlation to geometry, providing direct and simple way for corrosion mitigation.

Shell International E&P 2017 AIChE Annual Meeting 11

## RESULTS AND DISCUSSIONS – CONT'D

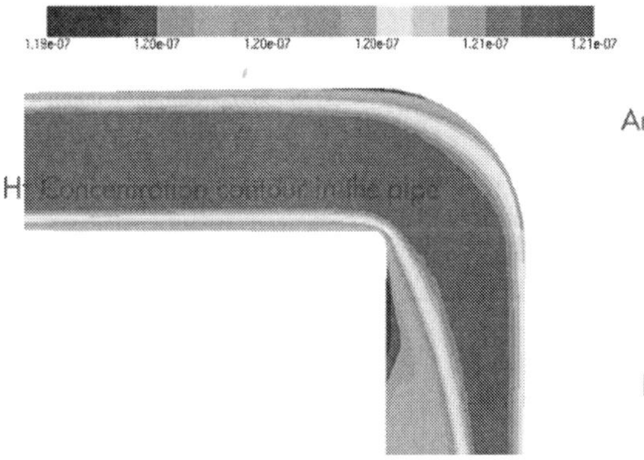


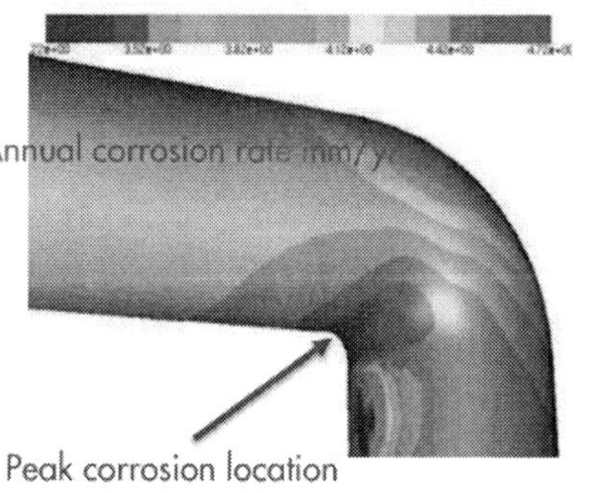


- Model can directly shows corrosion variations on surfaces and its correlation to geometry, providing direct and simple way for corrosion mitigation.
- Model can also shows whether the corrosion is flow assisted or not and provide better understanding for preventive measures.

Shell International E&P 2017 AIChE Annual Meeting 10/24/2017 12

## CONCLUSIONS

- CFD combined with corrosion chemistry can be a powerful tool for understanding corrosion mechanisms.
- Corrosion variations due to geometry and flow can be modeled and provide a more accurate way for preventive measures.
- Understanding of solution chemistry is also a key and can be added later to provide better understanding of the corrosion kinetics.
- Better mass transfer correlations may be available, further study may be worthwhile.
- Other corrosion types ($O_2$ and $H_2S$) can be implemented as well providing the corrosion and solution chemistry can be satisfactorily understood.

Shell International E&P 2017 AIChE Annual Meeting 10/24/2017 13

## ACKNOWLEDGEMENTS

- Dr. Yougui Zheng (Shell) for providing the corrosion kinetics information.
- Dr. Matthew Christiansen (Shell) for helping on the solution equilibrium chemistry
- Drs. Yi Dai (ANSYS) for providing the UDF foundation for the electrochemistry module in Fluent
- Shell's R&D support from Fluid Flow & Reactor Engineering

Shell International E&P 2017 AIChE Annual Meeting 10/24/2017 14

# 499856 Current State-of-the-Art of Gas Hydrates in Flow Assurance

**Abstract:**

This paper will review the current understanding and challenges of managing gas hydrates in flow assurance applications. The paper will be based largely on the developments performed over the last decade in the Colorado School of Mines, Center for Hydrate Research, and will also draw on the reported field and industry needs, challenges, and developments. The mechanisms of gas hydrate plug formation in different pipeline scenarios, including oil, water, and gas-dominated systems will be reviewed. This mechanistic understanding is based on extensive multi-scale experimental measurements, including flow loops, bench scale stirred cells, and interfacial and particle-level apparatuses. The different hydrate blockage mitigation strategies of hydrate avoidance with thermodynamic inhibitors, low dosage hydrate inhibitors, and exploiting the knowledge of hydrate kinetics will be assessed. This fundamental understanding is being incorporated into engineering tools that can enable the flow assurance engineer to predict if, when, and where hydrates will form and to formulate possible mitigation strategies.
Acknowledgements: Funding and support is acknowledged from the CSM Hydrate Center Industry Consortium and Coors Foundation (cak). Special thanks to the CSM Hydrate Busters, the talented students, postdocs and staff in the Hydrate Center.

# Case Study on Microbiological Control Program for Hydraulic Fracturing Operations Using Nitrate-Reducing Bacteria

Kiran Gawas and Chris Rodriguez

Halliburton



HALLIBURTON

## Hydraulic Fracturing—Microbial Contamination

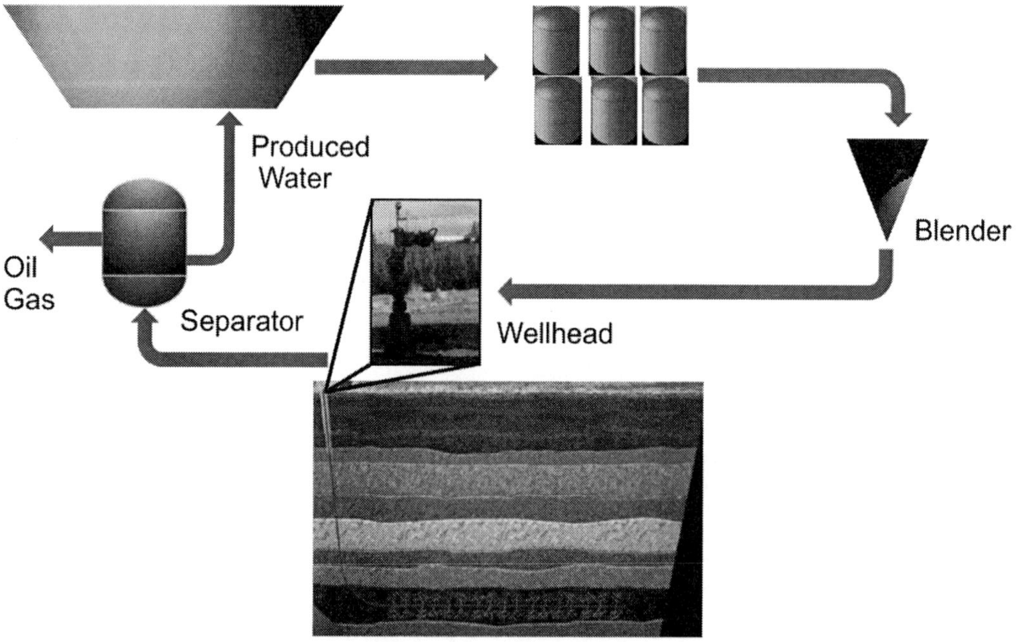

# Microbial Control and Asset Integrity

- Microorganisms encountered in oilfield application: sulfate-reducing bacteria (SRB), acid-producing bacteria (APB), slime, etc.

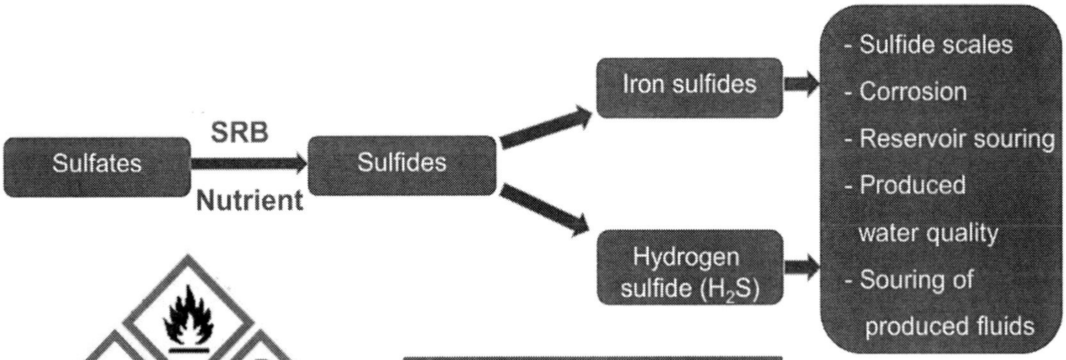

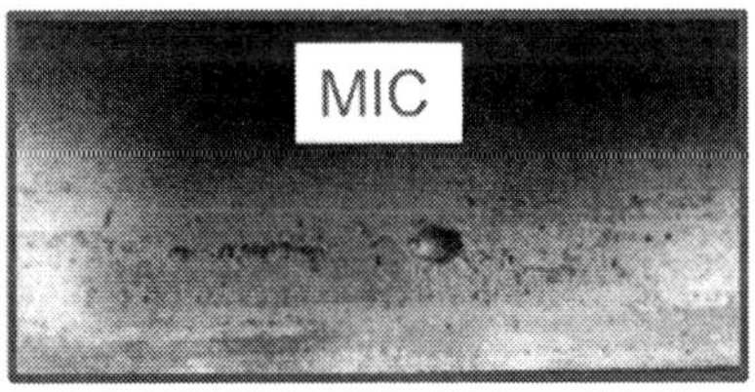


# Conventional Biocides

- Efficacy depends on pH, temperature, bacteria strains, speed of kill vs. half-life, fracturing chemicals, etc.
- Tolerance limits—aquatic organisms

| | $LC_{50}$ (PPM) | |
|---|---|---|
| Chemical | *Oncorhynchus mykiss* | *Daphnia magna* |
| Sodium hypochlorite * | 0.07 | 0.07 to 0.70 |
| ADBAC* | 0.93 | 0.025 |
| DBNPA * | 3.60 | 2.50 |
| Glutaraldehyde * | 13.0 | 5.00 |
| Dazomet | 16.0 | 11.0 |
| THPS * | 94.0 to 119.0 | 15.0 to 19.0 |
| Dimethyl oxazolidine (DMO) * | 93.0 | 48.0 to 51.0 |
| Sodium nitrate (40 %) | ➢ 1,000 (Oncorhynchus tshawytscha) | > 1,000 |

*Adapted from Williams and Cooper (2014) and Kahrilas et al. (2014). **Adapted from UCLID (2000).

- Potential environmental exposure
  - Surface spills—transportation
  - On-site spills—pits, leaks

# Conventional Biocides

- Efficacy depends on pH, temperature, bacteria strains, speed of kill vs. half-life, fracturing chemicals, etc.
- Tolerance limits—aquatic organisms

| | $LC_{50}$ (PPM) | |
|---|---|---|
| **Chemical** | ***Oncorhynchus mykiss*** | ***Daphnia magna*** |
| Sodium hypochlorite * | 0.07 | 0.07 to 0.70 |
| ADBAC* | 0.93 | 0.025 |
| DBNPA * | 3.60 | 2.50 |
| Glutaraldehyde * | 13.0 | 5.00 |
| Dazomet | 16.0 | 11.0 |
| THPS * | 94.0 to 119.0 | 15.0 to 19.0 |
| Dimethyl oxazolidine (DMO) * | 93.0 | 48.0 to 51.0 |
| Sodium nitrate (40 %) | ➢ 1,000 (Oncorhynchus tshawytscha) | > 1,000 |

*Adapted from Williams and Cooper (2014) and Kahrilas et al. (2014). **Adapted from UCLID (2000).

- Potential environmental exposure
  - Surface spills—transportation
  - On-site spills—pits, leaks

# Nitrate Reducing Bacteria (NRB) and Nitrate Control

- Environmentally acceptable alternative to conventional biocide treatment of SRB control in fracturing operations.


| | $LC_{50}$ (PPM) | |
|---|---|---|
| **Chemical** | ***Oncorhynchus mykiss*** | ***Daphnia magna*** |
| Sodium hypochlorite * | 0.07 | 0.07 to 0.70 |
| ADBAC* | 0.93 | 0.025 |
| DBNPA * | 3.60 | 2.50 |
| Glutaraldehyde * | 13.0 | 5.00 |
| Dazomet | 16.0 | 11.0 |
| THPS * | 94.0 to 119.0 | 15.0 to 19.0 |
| Dimethyl oxazolidine (DMO) * | 93.0 | 48.0 to 51.0 |
| Sodium nitrate (40 %) | ➢ 1,000 (Oncorhynchus tshawytscha) | > 1,000 |

*Adapted from Williams and Cooper (2014) and Kahrilas et al. (2014). **Adapted from UCLID (2000).


- Competitive exclusion of SRB using NRB live strain bio-nutrient matrix

## Case History: Biocides vs. NRB/Nitrate

- Location: Marcellus shale formation
- Relatively dry (low connate water content) and low porosity
- Treatment in more than 1,000 wells
- Temperature: 45 to 60°C
- Salinities: 100 to 150,000 PPM total dissolved solids (TDS)

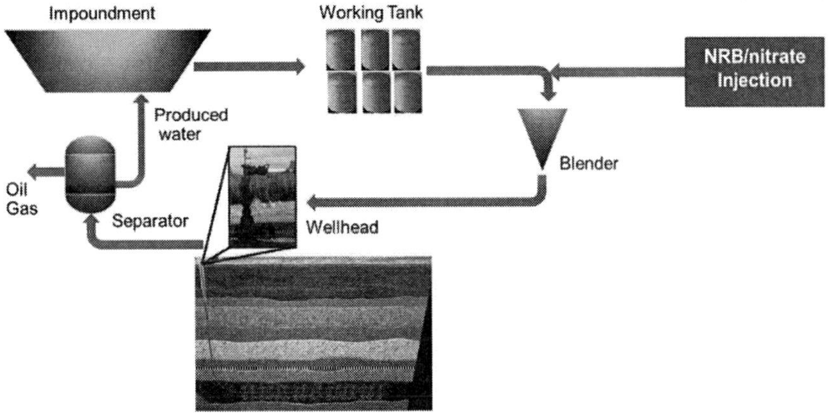


## NRB/Nitrate Mechanism

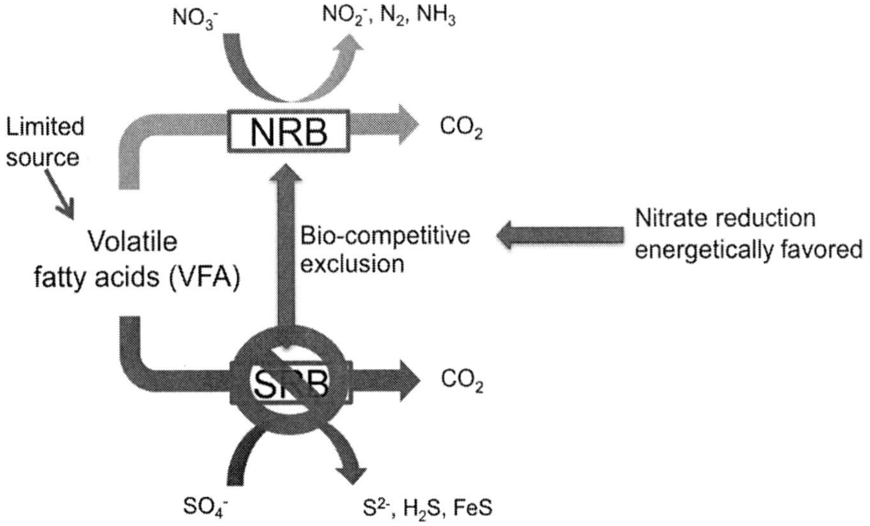


- Other mechanisms
  - Metabolic inhibition through production of nitrites
  - Direct oxidation of sulfide

## Case Study: Initial Results

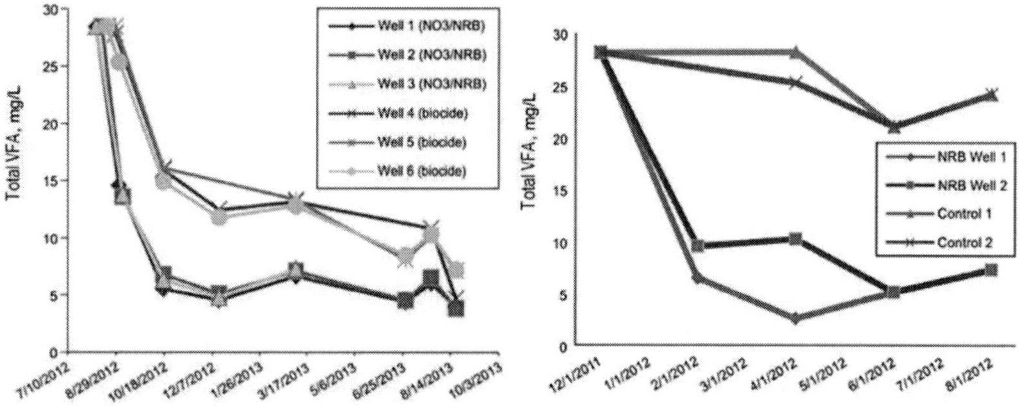


**VFA testing—NRB/nitrate vs. biocide**
[*Corrin et al. (2014), SPE 170937]

- VFA concentration reduced over time for selected wells.
- SRB population below detectable limits.
- Gas-phase sulfide at or below levels compared to biocide-treated counterparts

## Extended Case Study Results

- NRB/nitrate treatment program extended to more than 1,000 wells.
- Monitoring for SRB from 2010 to 2016.
- Wide range of salinities : 100 to 150,000 PPM TDS.
- No remedial well treatment for sulfides, biofilms or SRBs was required.

| | Total No. of Samples | No. of Samples Fouled | % Fouled |
|---|---|---|---|
| Operator 1 | 696 | 15 | 2.1 |
| Operator 2 | 93 | Below detectable limit | N/A |

## Summary

- Overall results show NRB/nitrate is as effective as a biocide.
- SRB was not recovered at the wellhead.
- NRB wells showed reduced $H_2S$ compared to biocide-treated wells.
- Sulfide scavengers application avoided.
- Use of conventional biocides avoided—reduced risk of potential personnel exposure.

## References


- Corrin, E., Harless, M., Rodriguez, C., Degner, D., and Archibeque, R., *"Evaluation of a More Environmentally Sensitive Approach to Microbiological Control Programs for Hydraulic Fracturing Operations in the Marcellus Shale Using a Nitrate-Reducing Bacteria and Nitrate-Based Treatment System"*, SPE Annual Technical Conference and Exhibition, Amsterdam, The Netherlands, 2014.
- Hubert, C., Voordouw, G., and Mayer, B., *"Elucidating microbial processes in nitrate- and sulfate-reducing systems using sulfur and oxygen isotope ratios: The example of oil reservoir souring control"*, Geochimica et. Cosmochimica Acta, 73 (13), 2009.
- Kahrilas, G. A., Blotevoge, J., Stewart, P. S., and Borch, T., *"Biocides in Hydraulic Fracturing Fluids: A Critical Review of Their Usage, Mobility, Degradation, and Toxicity"*, Environ. Sci. Technol., 2015, 49(1), 2014.
- Williams, T.M., and Cooper, L.E., "The Environmental Fate of Oil and Gas Biocides: A Review.", NACE CORROSION 2014, San Antonio, TX, USA, 2014.

THANK YOU

# 505765 Quantifying Gas Evolution Rates in Complex Hydrocarbon Systems

**Abstract:**

While much of gas-liquid separation is considered a mature technology, key challenges remain in designing compact systems â€“ particularly for subsea applications â€“ and heavy oil service. Separator performance is critically important for the safe and reliable operation of downstream equipment. In compact separators or heavy oil service, the rates of gas evolution are particularly important design parameters. Incomplete removal of gas from liquid, or gas carry-under, may result in unexpected pump cavitation, hydrate formation, erosion, and slugging. In petroleum emulsions, gas evolution can actually impact the emulsion morphology, thereby affecting subsequent oil/water separation. Conventional separator design typically only accounts for the entrained gas bubbles in the liquid, in addition to bulk separation. Therefore, the kinetics of solution gas evolution is a critical uncertainty in gas-liquid separator design, especially in systems that contain surfactants, dissolved solids, and other complex species. This presentation provides a discussion about recent gas evolution data collected in hydrocarbon systems with special focus on petroleum emulsions.

# 505764 Sand Particle Erosion in Single-Phase and Multiphase Flow

**Abstract:**

Sand particle erosion modeling is a challenging task for engineers specially when it comes to gas-liquid-sand multiphase flow where different interfacial structures are present in the flow and the flow demonstrates highly-transient behavior.

The current work reviews available models and approaches for modeling erosion in single-phase and multiphase flow. Specifically, a comprehensive and advanced approach for investigating multiphase flow and erosion in multiphase flow is presented where capability of combining CFD and Wire-Mesh Sensor techniques is shown. Moreover, erosion prediction mechanistic models and empirical correlations are demonstrated. Finally, areas requiring more attention for further work are indicated.

# 491076 Study on the Pig Motion in Waxy Crude Oil Pipeline While Wax Removing Operation

**Abstract:**

Wax precipitates when waxy oil temperature decreases in pipeline transportation. A section of slurry (mixture of wax deposits and oil flow) may come into being when pigging is carried out for removing wax deposit, which can even lead to the getting stuck of a pig. So, it is necessary to find reliable tools to predict many variables related to the motion of a pig, such as the pig velocity, the pressure profile along the pipeline and the bypass flow rate, which are important to pigging efficiency and safety.

Reliable predictions of above variables demand appropriate models to handle (i) basic problems like the friction force between pipeline wall and pig, the resistant force of deposits and parameters of slurry formed in front of pig as well as (ii) the solution to combine the motion of pig, the motion of fluids and the motion of slurry.

To tackle these two problems, a mathematical model was developed to simulate the motion of a pig in waxy crude oil pipeline. In contrast with prior studies in which pigs moved in liquid and gas pipelines, wax deposits attached to pipeline wall was taken into consideration for the first time. So in addition with the fluid flow equations and dynamic equations of pig where key parameters like wax resistant force and friction force were obtained through experiments, a model to simulate the increasing slurry in front of pig was also established. All of them were solved together using the method of characteristics. Note that interaction between pig and fluid flow (or slurry) was described by mass conservation equations at boundaries. Combining pig motion with the pump operating characteristics, pig velocity and pipeline inlet pressure under any conditions of wax thickness can be obtained through the model (figure 1). Wax resistant force turns out to be two orders of magnitude smaller than pressure drop of the slurry section even when the thickness of wax deposits is 0.1m. Bigger diameter of bypass hole in the pig can, on one hand, reduce the viscosity of slurry which benefits the pig motion, while decrease pig velocity on the other hand, so an optimal diameter of bypass hole is recommended for the sake of pigging time consumed. According to the calculation results, when bypass diameter increases from 0.01m to 0.03m, the pressure drop of slurry decreases by nearly 15% after the pig moved for the same time.

Figure 1 Time evolution of pig velocity and pipeline inlet pressure under different thickness of wax deposits ($h$)

Pigs would get stuck once the pipeline inlet pressure exceeds the available pumped pressure, so pipeline inlet pressure was taken as a criterion to judge whether a pig would get stuck (figure 2). Finally, sensitivity analysis in view of the effect of related parameters on the increasing rate of pipeline inlet pressure was performed and the results indicated an increasing sequence: thickness of wax deposits, friction force between pipeline wall and pig, viscosity of slurry.

Figure 2 Illustration of pressure drop along the pipeline during pigging at different times (the red sign represents the pig position)

# 505769 Sand Agglomeration in Oil & Gas Reservoirs using Polymers

**Abstract:**

Sand production is an age-old problem faced by the oil and gas industry which can have severe cost implications if not controlled properly. Approximately 70% of the worldâ€™s remaining hydrocarbons are located in sand-prone reservoirs and current approaches to handle sand production are still facing many challenges. The aim of this work is to explore the use of polyacrylamide-based polymers in agglomerating formation sand and clays to improve the effectiveness of standalone sand screens. Polymer bridging interactions, charge neutralization and bare patch attraction are the three main agglomeration mechanisms being considered due to their capability of forming aggregates with superior shear resistivity and size. Preliminary results using Scanning Electron Microscopy (SEM) and Laser Particle Size Analysis (LPSA) have indicated a median growth in particle size of around 5 times when treated with cationic polyacrylamides only. Multi-polymer agglomeration systems were able to produce sand agglomerates of similar sizes but with much higher resistance to shear and more able to retain its structure upon drying. The novelty of this multi-polymer agglomeration system when coupled with sand screens is in the ability to potentially provide a very cost-effective means to solve sand production issues at the source, eliminating the need for costly remedial solutions at the topside and resin-based sand consolidation methods which impair reservoir permeability.

# 494704 Characterization of Hydrate Slurry Transportability: Comparison of High Pressure Rheometer Measurements with Industrial Scale Flowloop Data

**Abstract:**

**Characterization of Hydrate Slurry Transportability: Comparison of High Pressure Rheometer Measurements with Industrial Scale Flowloop Data**

*Ahmad AA Majid*[1,2], *Carolyn A Koh*[1*]

1. Center for Hydrate Research, Colorado School of Mines, Golden CO 80401, USA
2. Faculty of Chemical & Natural Resources Engineering, Universiti Malaysia Pahang, Malaysia

## ABSTRACT

Gas hydrates are solid inclusion compounds that contain small gas molecules trapped inside a lattice of water cages, and typically form at high pressure and low temperature. In the oil and gas industry, the formation of gas hydrates in flowlines is a major concern as it can cause plugging of the flowlines. This is a major concern as it poses both safety and environmental risks. The transportability of a gas hydrate slurry was evaluated using two different apparatuses; both large and lab-scale measurements. For the large-scale measurements, an industrial-scale flowloop was used and the transportability of the hydrate slurry was evaluated from pressure drop data (see (Grasso et al. 2014; Majid et al. 2016)). In lab-scale measurements, in-situ viscosity measurements of hydrate slurries were conducted using a high-pressure rheometer. Both apparatuses have advantages and disadvantages. In order to compare the results across both apparatuses, experimental parameters were specifically chosen to ensure both systems have relatively similar conditions. Two different parameters were investigated in this work; water cut and shear rate. Specifically, three different water cuts (30, 50 and 90 vol.% water cut) and five different shear rates were investigated for a crude oil system. In this investigation, there is qualitative agreement between results from the rheometer with the flowloop. From the analysis of the gas consumption data at similar experimental conditions (water cut and pressure), similar amounts of hydrate formed in both the rheometer and flowloop. Furthermore, in order to qualitatively compare the results from the rheometer with the results obtained from the flowloop, relative calculations (relative pressure drop and relative viscosity) were used. Results from the relative calculations show that the relative viscosity and relative pressure drop are on a similar order of magnitude. However, relative viscosity is higher than the relative pressure drop. Finally, in this work, the relationship between relative viscosity and relative pressure drop was developed investigated. This relationship can ultimately help flow assurance engineers in evaluating the transportability of hydrate slurries in flowlines.

**References:**

Grasso GA, Lafond PG, Aman ZM, et al. 2014. Hydrate Formation Flowloop Experiments. In: The 8th International Conference of Gas Hydrates.

Majid AA, Lee W, Srivastava V, et al. 2016. The Study of Gas Hydrate Formation and Particle Transportability Using a High Pressure Flowloop. In: Offshore Technology Conference.

# 496254 Alkylammonium Formate-Based Protic Ionic Liquids for Methane Hydrate Inhibition: Offshore Flow Assurance

**Abstract:**

**Alkylammonium Formate-Based Protic Ionic Liquids for Methane Hydrate Inhibition: Offshore Flow Assurance**

T. Altamash, F. Qureshi, M. Atilhan, M. Tariq and M. Khraisheh*

**Department of Chemical Engineering, Qatar University**

**Author Emails**: tausif.Altamash@qu.edu.qa|mQureshi@qu.edu.qa | mert.atilhan@gmail.com

***Author for Correspondence*:*** *m.Khraisheh@qu.edu.qa*

**Abstract**

During the transportation of natural gas via pipeline on the sea floor where high pressure and low temperature exist a severe flow assurance problem arises due to formation of gas hydrates. Gas hydrates are solid crystalline compounds that are formed under favorable thermodynamic condition. Practically, the formation of hydrates is a constant threat to the oil and gas industry, as they have the tendency to grow larger in sizes and completely block the pipeline over the period of the time. In addition to that, they can cause major operations shutdown and an explosion leading to large economic losses and posting safety risk to nearby operators.

In order to solve hydrate formation issue some conventional thermodynamic inhibitors like Methanol and Methyl ethylene glycol (MEG) are widely used in industry. However, these inhibitors are highly flammable, toxic in nature required in bulk quantities and leakage may disturb the aquatic environment. Therefore, there is a high demand for new class of low dosage somehow greener gas hydrate inhibitors to keep away from hazards.

Ionic liquids (ILs) are asymmetric structure containing salts that are also known as designer solvents due to their ability to be tailor according to the process requirement. Some classes of ILs are also reported to be environmentally friendly and non-flammable in nature. Generally, the ILs are classified as Aprotic and Protic ILs. The protic ionic liquids are formed by proton transfer from Bronsted acid to Bronsted base. In comparison to Aprotic, the protic ILs are easier methodology and one step reaction process made this class of IL cheaper in cost. Therefore, five different simpler protic ILs have been selected as; Ethyl ammonium formate, Dimethyl ammonium formate , Dimethylethyl ammonium formate, Tributyl ammonium formate, and were tested with methane gas at different pressures and concentrations to find the effect of high dosage by PSL systemtecknik rocking cell assembly.

According to the experimental results, at lower concentration (5wt%) the selected protic ILs act as kinetic hydrate inhibitors. But at higher concentration (10wt%) the selected protic ILs exhibit both thermodynamic and kinetic inhibition effect simultaneously. The methane HVLE curve was shifted by $2.0 \pm 0.1$ °C and hydrate formation was delayed by 2 hrs at high pressure conditions (~120 bars). The kinetic inhibition performance of protic ILs was also significantly enhanced when they were doped with synergents and the hydrate was delayed upto 20 hrs at low pressure conditions (~40 bars) and 10 hrs at high pressure conditions (~120 bars). Moreover, poly ethylene oxide (PEO) were added at lower amount (1wt%) with 5 wt% of said protic ionic liquids. As expected the PEO shown enhanced kinetic effect whereas, no thermodynamic effects were observed.

**Acknowledgement**

*This work was made possible by NPRP grant for # 6-330-2-140 (TA) and GSRA # 2-1-0603-14012 (FQ) for from the Qatar National Research Fund (a member of Qatar Foundation). The statements made herein are solely the responsibility of the authors.*

# 496805 Solid Nanoparticles As Hydrate Inhibitors

**Abstract:**

Clathrate hydrates are non-stoichiometric, ice-like crystalline solids that can lead to plugging of crude oil pipelines. A large amount of money is spent by the petroleum industries to mitigate hydrate formation. In an oil-dominated system, emulsion stability is an important criterion for hydrate risk management. The extent to which the emulsion droplets remain segregated during hydrate formation/dissociation determines the level of hydrate plug formation. We have investigated the potential of solid nanoparticles of different wettabilities to potentially mitigate hydrate formation in surfactant stabilized water-in-oil emulsions. The primary objective is to elucidate the change in interfacial properties of surfactant stabilized water-in-oil emulsions upon addition of solid particles of different wettabilities and the resulting consequences on hydrate formation and flow behavior.

A DHR-3 stress controlled rheometer whose temperature can be controlled between -20°C to 150°C is used to examine the rheological behavior of hydrate forming emulsions. A helical ribbon geometry will be used to study the rheological behavior of hydrates. An Olympus BX53 polarized microscope equipped with a high-speed camera and Linkam shear- and temperature-controlled stage will be used to characterize the droplet size and wax crystals in water-in-emulsions. DSC will be used to characterize the emulsion stability, hydrate formation and wax formation in water-in-oil emulsions. Cyclopentane hydrates are studied in model oil systems using surfactant and silica nanoparticles, which act as stabilizing agents.

The results show that the addition of silica nanoparticles of different wettabilities to a surfactant stabilized emulsion significantly alters the interfacial properties of water-in-oil emulsions, which in turn had a profound effect on hydrate formation and flow behavior of hydrate forming water-in-oil emulsions.

# 501140 Asphaltene Precipitation from a Brazilian Crude Oil By $CO_2$ Injection or By Pressure Reduction

**Abstract:**

Asphaltenes constitute a petroleum fraction characterized by low solubility in n-alkanes and high solubility in aromatic compounds. In general, such compounds are molecularly solubilized in the oil or are present in a colloidal dispersion. Changes in the temperature, pressure and composition of the oil may lead to the precipitation of these compounds which can decrease or even interrupt the oil production. The natural gas associated with oils from the Santos basin presents high levels of $CO_2$. These high levels do not allow the delivery of natural gas for use on land, since the maximum $CO_2$ content in this stream is 5%. In order to enable the production and maintenance of reservoir pressure, natural gas is fractionated in membranes and the $CO_2$-rich fraction is reinjected into the reservoir. Thus, precipitation of asphaltenes may occur during the production of the oil caused by the decrease in both the temperature and the pressure of the oil or in the injection well where a variation in the composition of the system occurs. In this work, experimental results are presented for the determination of the asphaltene precipitation onset of the reconstituted oil with natural gas both by the isothermal reduction of the pressure and by the injection of a mixture of $CO_2$ and methane. For determination of these onsets, a solids detection system composed by a laser at 1550 nm and a transmitted light power detector was assembled on a commercial variable volume cell. To verify the presence of precipitated material, the isobaric filtration of part of the oil was performed using a filter with a cutoff diameter of 0.5 µm. In a preliminary experiment it was observed that there was asphaltene precipitation even with small amounts of $CO_2$-methane injected.

ACKNOWLEDGEMENTS
The authors acknowledge the support of ANP – Agência Nacional de Petróleo, Gás Natural e Biocombustíveis and Petrogal Brasil S.A., related to the grant from R&D investment rule.

# 493505 Effect of 1-Pentanol on Wettability of Oil/Brine/Rock Systems

**Abstract:**

In this article, we evaluated the potential of applying a medium-chain alcohol like 1-Pentanol in low-salinity waterflooding. We reported for the first time the contact angle of a variety of Oil/Brine/Rock systems when 1-Pentanol was present or absent in the brine solution. Adding 1 wt% 1-Pentanol to the brine only produced a modest effect on the contact angle of an oil from a sandstone reservoir on sandstone-like substrates (i.e., mica and quartz) in NaCl and $MgCl_2$ solutions over concentrations of 0-3 M. We also observed that adding 1 wt% 1-Pentanol to the brine could significantly lower the contact angle of an oil from a sandstone reservoir on carbonate-like substrates (i.e., calcite) in NaCl and $MgCl_2$ solutions over concentrations of 0-1 M. Notably, the contact angle of the Oil/NaCl/Calcite system at 0.1M NaCl was reduced by ~80° after 0.5 wt% 1-Pentanol was added to the brine, indicating that the surface became strongly water-wet which may favor oil recovery. The zeta potential of rock/brine and oil/brine interfaces were also measured and discussed to understand the trends observed in contact angle. These findings may provide new perspectives and insights to low-salinity waterflooding.

# 486724 Flow Assurance Issues Associated with Ice Deposition Under Freezing Conditions


**Abstract:**

## Abstract

With the ever increasing global energy needs, oil companies have taken aggressive measures to ensure continuity in oil supplies. This has led to the construction of oil pipelines in harsh environments including extremely low temperatures. Under these temperature conditions, water even in trace amounts may form ice and lead to pipeline blockages and associated risks. A pipeline blockage due to an ice plug was reported to delay the restart of the Poplar pipeline system gathering crude oil from Montana and North Dakota. Ice formation in oil pipeline has been considered a high risk threat by Trans-Alyeska Pipeline System leading the company to invest millions of dollars into ice formation research. The declining throughput makes the oil cool to lower temperatures. In oil temperature falls below the freezing point, ice forms which may lead to valve failure, flow restriction due to ice accumulation, and even plugging of pipelines.

This paper presents low water cut (<2 %) ice formation studies and its effects on the oil transportation system. A 2-inch diameter carbon steel flow loop was instrumented to measure droplet size, pressure, temperature, and differential pressure. The water cut and wax content changes with time were monitored. Three test fluids were used, including kerosene (40° API, 1.6cp at 70 °F), a light crude oil (40° API, 2% wax content, 11cp at 70 °F), and a blend of the two oils. The effect of pipeline geometry was studied, including low spots, straight pipe section with an inserted rod, and straight pipe section with a perforated plate. Particle interactions, crystallization, and ice melting process were observed using Particle Video Measurement (PVM).

Results indicated that, like wax, ice deposition requires a negative temperature gradient. Visual observations from PVM showed that ice deposition resulted from either freezing water in low spots or from ice particles migrating from the bulk to the wall. Flowline plugging resulted from excessive accumulation of ice particles or the wall and/or from severe aggregation of ice particles that eventually get stuck in flow restrictions such as perforated plates or inserted metal rod. The morphology of the deposited ice strongly depends on oil composition; the oil with 2% wax resulted in glazed ice whereas the kerosene, no wax, resulted in rime ice. The morphology of ice is very important since it defines the hardness and other properties of ice. Additionally, amount of the deposited ice increased with increasing wax content. This suggests that the porosity of ice was affected by presence of wax, i.e., the interaction of ice-wax affects the amount of the occluded hydrocarbons. Furthermore, the hydrocarbons in deposit contained a higher wax fraction than the test oil. The oil with 2% wax resulted in a delayed onset but quick plugging behavior whereas kerosene resulted in quick onset but gradual plugging behavior. Other test conditions and fluid properties affecting ice deposition were also studied such as flow geometry, watercut and flow rate. Lastly, CFD simulations were performed for the different geometries and results showed qualitatively agreement with experimental results.

# 490221 Electrical Treatment of Waxy Crude Oil to Improve Low Temperature Flowability

**Abstract:**

Wax precipitation from crude oils caused by temperature reduction greatly increases the apparent viscosity of oils and thereby poses challenges to the safety and efficiency of their transportation. In the present work, we demonstrate that the apparent viscosity of waxy crude oil at temperatures near the pour point could be significantly reduced by applying a direct electric field parallel with the direction of the oil flow for only one to two minutes. A novel apparatus (Figure 1) was assembled that can electrically treat waxy crude oil and simultaneously measure the rheological property of the treated oil, and a method was developed to calculate the viscosity variation by using non-Newtonian fluid mechanics and rheological principles. Lower treatment temperatures (Figure 2) and higher electric field strengths (Figure 3) provided greater viscosity reduction. We also observed that the performance of electrical treatment depends upon the shear rates, with a lower shear rate generally resulting in a greater viscosity reduction. Notably, a viscosity reduction of 70% was obtained after the crude oil, that has a pour point of 17 °C, was electrically treated at 18 °C and 0.8 kV/mm for 90 s. Besides, the energy consumption of the electrical treatment was estimated to be less than 1% of that of the conventional heating method to achieve the same viscosity reduction performance. Thus, we herein demonstrate a technology for improving the flowability of waxy crude oils at low temperatures in an energy-efficient manner.

Figure 1: Schematic diagram of the electrical treatment apparatus

Figure 2: Effect of treatment temperature on viscosity reduction at different shear rates

Figure 3: Effect of electric field strength on viscosity reduction at different shear rates

# 491279 A Transient, Two Fluid Model for Slug Flow Characterization

**Abstract:**

In this work, we propose a numerical resolution of a 1D, transient, simplified two-fluid model regularized with an artificial diffusion term to model stratified, wavy, and slug flow in horizontal and nearly horizontal pipes.

The final aim is to provide transient model to predict slug formation and evolution that can be used in pipelines.

The artificial diffusion is adopted to prevent the unbounded growth of instabilities outside the confines of the conditions under which the one-dimensional transient two-fluid model is well-posed; the artificial diffusion values are set case by case to obtain the desired cut-off at short wavelengths: the choice is based on the analysis of the linear stability results. A proper criterion to simulate two-phase to single phase flow transition, which occurs during slug formation, was developed. Flow patterns transitions have been numerically computed and compared against theoretical transition boundaries and experimental observations. Moreover, we showed that the code computes slug initiation and slug characteristics, in a reasonable accurate way considering the simplicity of the model, comparing numerical results with well-known empirical relations and experimental data.

Geometries involving slope changes and different inclinations are tested.

Furthermore, the model simplicity leads to a computationally cheap numerical resolution; this can be useful in engineering applications where obtaining fast numerical results is fundamental, such as application involving automated control for two-phase flows

# Three-Phase Equilibrium Computations for Hydrocarbon-Water Mixtures using a Reduced Method

Michael Connolly, Huanquan Pan, Hamdi Tchelepi
Department of Energy Resources Engineering, Stanford University

## Introduction

Thermal recovery is the most ubiquitous and successful EOR method. Techniques include steam-flooding, cyclic steam injection and steam-assisted-gravity-drainage (SAGD).

Phase behavior in thermal compositional simulation has traditionally been addressed using pre-computed equilibrium ratios (K-values). However, K-values have no intrinsic thermodynamic content, and thus do not provide a consistent basis from which to compute thermodynamic properties of a fluid phase. A full equation-of-state (EOS)approach is more rigorous, though remains prohibitively costly.

## Research objectives

Design and implement a phase equilibrium kernel to answer two questions:
1. How many phases are present?
2. What is the composition and amount of each phase?

We focus on phase equilibria of hydrocarbon-water mixtures in the context of thermal reservoir simulation. Our objective is twofold:
a) Account for all components in all phases.
b) Convergence of phase split to tight tolerance for reservoir simulation.

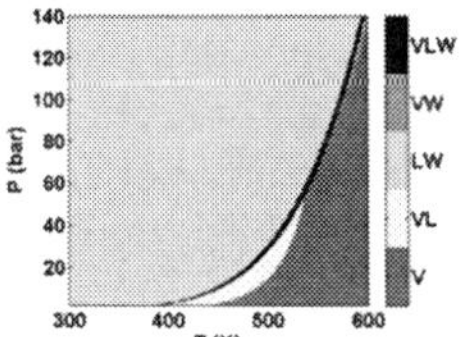


PT phase diagram for 95% water/5% oil mixture. Narrow 2 & 3-phase regions are present

## Methodology

There are two broad approaches to phase equilibrium calculations in reservoir simulation
I. Pre-allocate the number of phases, estimate compositions and proceed directly to the phase-split calculation.
II. Perform stability-testing and phase-split calculations in a step-wise procedure.

We choose the latter method, in the interests of developing a rigorous solution strategy. Our approach involves:
1. Stability testing, tailored to the distinctive characteristics of hydrocarbon-water mixtures.
2. A reduced method for phase split calculations.

## Phase stability testing

- Phase stability testing can be deceptively difficult for hydrocarbon water mixtures at elevated temperatures. Narrow regions of immiscibility and multiple two-phase states occur.

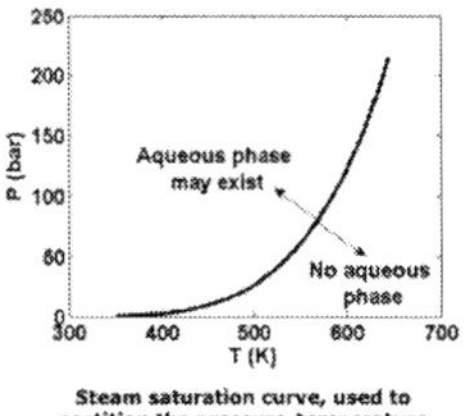


Steam saturation curve, used to partition the pressure-temperature domain for stability testing

To initialize stability tests we estimate K-values by partitioning the pressure-temperature domain using the steam saturation curve. Our rationale is that the aqueous phase cannot exist below the steam saturation pressure. Above the steam saturation curve the aqueous phase may be present, along with a vapor phase and an oleic phase.

$$K^{Wilson} = \frac{P_c}{P}\exp\left[5.37(1+\omega_i)\left[1-\frac{T_c}{T}\right]\right]$$

$$K^{Aq}_{H_2O} = \frac{0.999}{z_{H_2O}} \ \& \ K^{Aq}_i = \frac{0.001}{(N_c-1)z_{H_2O}}$$

Initial estimates

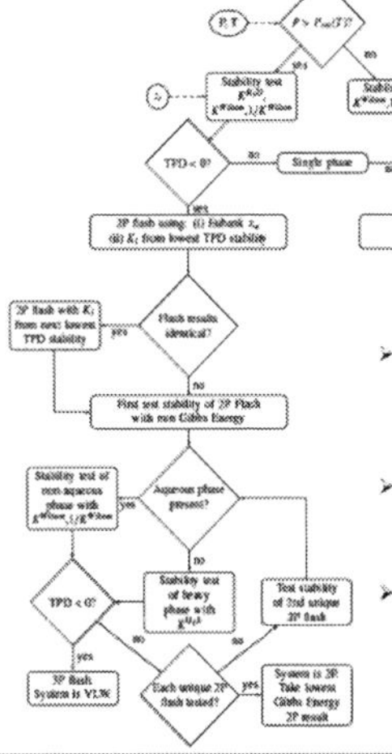

- Stability testing provides initial estimates for phase split calculations
- Multiple two-phase split calculations are required
- Composition yielding the lowest TPD does not always lead to the correct flash result

## Reduced method for flash calculations

- We develop a reduced formulation for three-phase split calculations. We use the Peng Robinson cubic EOS, which consists of energetic and co-volume terms:

$$P = \frac{RT}{\bar{v}-b} - \frac{\alpha a}{\bar{v}+2b\bar{v}-b^2} \qquad a = \sum^{c}\sum^{c} x_i x_j \sqrt{a_i a_j}\,(1-\delta_{ij})$$

- Component interactions are manifest in the matrix of binary-interaction-parameters, $\beta = (1-\sigma_{ij})$. We use the spectral decomposition to re-write $\beta$:

$$\beta = Q\Lambda Q^T = \sum_i \lambda_k q_{ki}' q_{kj}' \implies a = \sum_i \lambda_k Q_k^2$$

- Solution of a cubic EOS requires both the compressibility factor $\mathbf{Z}$ and fugacity coefficients $\hat{\phi}_i$, which can be written in terms of the energetic and co-volume parameters:

$$\begin{aligned} Z &= Z(Q_1, Q_2, \dots, Q_{Nc}, b) \\ \hat{\phi}_i &= \hat{\phi}_i(Q_1, Q_2, \dots, Q_{Nc}, b) \end{aligned} \quad \xrightarrow[\ m < N_c\ ]{\text{reduction}} \quad \begin{aligned} Z &= Z(Q_1, Q_2, \dots, Q_m, b) \\ \hat{\phi}_i &= \hat{\phi}_i(Q_1, Q_2, \dots, Q_m, b) \end{aligned}$$

- The set of reduced variables consists of
  1. Co-volume parameters, b
  2. Linear combinations of eigenvalues and eigenvectors of the BIP matrix, $Q_m$
  3. $N_p-1$ phase amounts ($N_p$ the number of phases)

- We implement Newton's method to solve the phase-split. The partial derivatives of the reduced variables are convoluted. We verify our implementation using Stanford's Automatic Differentiation Extensible Template Library (ADETL).

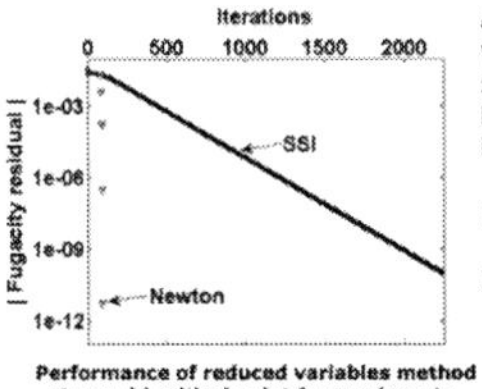


Performance of reduced variables method at near bi-critical point for condensate-water mixture. Successive substitution alone takes over 2400 iterations.

Advantages of a reduced variables approach to phase split calculations involving hydrocarbon-water mixtures include:

1. Smaller system of equations.
2. Full-rank linear system by construction, even in the presence of trace components.
3. Rapid convergence, even at bi-critical points.

## Testing & analysis

- We tested over an extensive parameter space by varying P, T and $\mathbf{z}$. We used 8 characterized fluids from the literature.

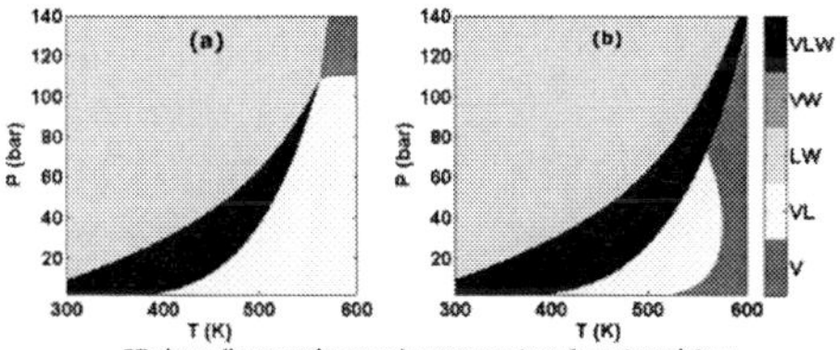


PT phase diagrams for pseudocomponent crude-water mixture
(a) 50% water/50% hydrocarbon (b) 99% water/1% hydrocarbon

## The link to thermal simulation

- Thermal recovery processes involve complex multiphase, multicomponent interactions in porous media. Water and hydrocarbons partition across multiple fluid phases as a function of pressure, temperature and composition.
- Water is the thermodynamically dominant component in thermal settings. Phase transitions are abrupt and occur frequently, owing to steep temperature gradients in the reservoir. Narrow boiling point behavior occurs as large enthalpy changes coincide with small temperature changes.
- There exists a highly nonlinear coupling between local thermodynamic constraints and the conservation laws at the global level. Rigorous quantification of phase behavior is essential to the fidelity of this coupling.

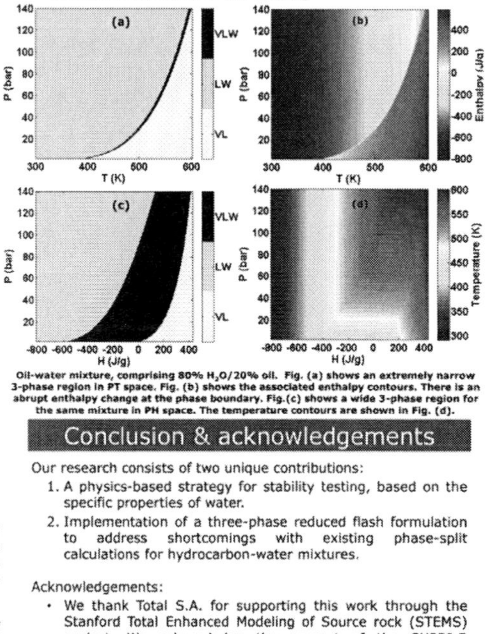


Oil-water mixture, comprising 80% $H_2O$/20% oil. Fig. (a) shows an extremely narrow 3-phase region in PT space. Fig. (b) shows the associated enthalpy contours. There is an abrupt enthalpy change at the phase boundary. Fig.(c) shows a wide 3-phase region for the same mixture in PH space. The temperature contours are shown in Fig. (d).

## Conclusion & acknowledgements

Our research consists of two unique contributions:
1. A physics-based strategy for stability testing, based on the specific properties of water.
2. Implementation of a three-phase reduced flash formulation to address shortcomings with existing phase-split calculations for hydrocarbon-water mixtures.

Acknowledgements:
- We thank Total S.A. for supporting this work through the Stanford Total Enhanced Modeling of Source rock (STEMS) project. We acknowledge the support of the SUPRI-B industrial affiliates.

## References

Pan, H., Firoozabadi, A., 2003. SPE Journal
Connolly, M., Pan, H., Tchelepi H.A., *In Review*, SPE Journal
Connolly, M., Pan, H., Tchelepi H.A., *In Preparation*, SPE Journal

# 499758 Artificial Intelligence Applications to Forecast Oil Production from Hydraulically Fractured Reservoirs

**Abstract:**

Applications of artificial intelligence (AI) are rapidly growing in every field from medical diagnosis to robot control due to this method's innovative techniques in connecting inputs to outputs and recent advancements in high-speed computing. The oil and gas industry has started exploring the possibilities of AI applications in various sectors such as reservoir characterization, fluid thermodynamic properties, production maximization, fracture job optimization, recoverable hydrocarbon estimation, well placement using pattern recognition, etc. In this study, two popular artificial intelligence techniques, namely the Least Square Support Vector Machine (LSSVM) and Artificial Neural Networks (ANN) methods are developed as surrogate models to forecast oil production from hydraulically fractured reservoirs such as shales and other tight formations. The objective of this work is to evaluate AI techniques as alternative methods for production performance analysis. Eight important factors for production such as reservoir permeability, gas relative permeability exponent, rock compressibility, initial gas oil ratio, slope of gas oil ratio versus pressure, initial pressure, flowing bottom hole pressure and fracture spacing are selected as input for this study. The ranges of inputs are based on real field data from the Eagle Ford in Texas, Bakken in North Dakota and Niobrara in Colorado-Wyoming in the USA. Two important production performance measures, i.e. oil recovery and produced gas oil ratio (GOR) are selected as output of the models. Oil recovery and gas recovery are recorded at particular time intervals (90 days, 1 year, 5 years, 10 years, and 15 years) and after the oil rate drops to 5 barrels per day per fracture. Therefore, five time-based models and one rate-based model are developed using LSSVM and ANN for oil recovery and GOR. Surrogate models (LSSVM and ANN) are trained with 80% of data (training set) and are verified with 20% of data (test set) based on numerical simulation. Particle Swarm Optimization (PSO) is used to find the model parameters. Fitness of the models are evaluated by measuring the coefficient of determination and normalized root mean square error. LSSVM models have shown very high accuracy compared to ANN models. The complex nature of produced GOR are successfully predicted by LSSVM. These models can be used as proxy models for reservoir simulator to forecast recovery and to perform sensitivity and uncertainty analyses.

# 506114 Development of a New Model and Evaluation of Various Methods of Predicting Dew Point Pressure for Gas Condensate Reservoirs

**Abstract:**

Dew point pressure plays a critical role in the development of gas condensate reservoirs. Reserve estimation, production performance evaluation, and production systems design all are affected by dew point pressure. When reservoir pressure decreases below the dew point pressure, liquid drops out, and well productivity decreases rapidly. Therefore, estimation of dew point pressure is very essential for gas condensate reservoirs.

This paper is focused on the development of a new dew point pressure model. Data was collected from a number of different sources and subjected to a detailed filtration process. The data representing 715 gas condensate samples covering a wide range of gas compositions, at different pressures and reservoir temperatures. Industry methods for characterizing gas condensate based on total molecular weight and composition were tested and were found to be invalid. Theoretical and statistical analyses of the collected data were carried in order to develop a new model. This model relates the dew point pressure to gas composition, characterization of Heptane plus fraction, and reservoir temperature. It has an average absolute error, root mean square error, and coefficient of determination of 7.5%, 588, and 0.87, respectively. The proposed model has been compared to published correlations as well as an un-tuned equation of state, and it resulted in a better accuracy. The validity of the proposed model was tested against experimental data and equation of state simulations. It succeeded in capturing the physical trends associated with varying different input parameters.

# AUTHOR INDEX

# AUTHOR INDEX